Know Your Enemy

LATE-1944-1945

CONTENTS

Compiled by: Tony Vodanovich, Sean Goodison, Wayne Turner from work produced by the Battlefront team and friends
Photographs: National Library of New Zealand, Imperial War Museum, Bovington Tank Museum, Australian War Memorial, NARA, National Archives Of Canada, US Army Signal Corps

This is a complement to the *Flames Of War, the World War II miniatures game.*
A copy of the rulebook for *Flames Of War* is necessary to fully use the contents of this book.

USA

NATIONAL RULES

TANKS

Stabilisers

A moving tank fitted with a Stabiliser can fire its main gun at its full ROF, but adds a penalty of +1 to the score needed to hit.

Before shooting, a tank can choose not to use its Stabiliser and fire at the normal rate if that would give it a better chance.

Hedgerow Cutters

When a tank fitted with a Hedgerow Cutter attempts to cross a Bocage Hedgerow, it successfully crosses on a roll of 2+ rather than a Skill Test. If it fails, the tank Bogs Down on the Hedgerow as normal, but ignores the Belly Up rule (see page 58 of the rulebook).

If it crosses successfully, then any other Fully-tracked Tank teams that started their movement adjacent to the Hedgerow can follow it through on a roll of 2+ instead of a Skill Test as well.

TRANSPORTS

Dismounting Machine-guns

Any Infantry team carried as a Passenger in a Transport team may Dismount in the Movement Step with the vehicle's Passenger-fired .50 cal AA MG, becoming an M2 .50 cal team, or with its Passenger-fired AA MG becoming a LMG team. If the Infantry team was a Command team, it remains a Command team.

An Infantry team normally mounted in a vehicle can start the game Dismounted with their vehicle's Passenger-fired machine-gun.

Once the machine-gun has been Dismounted from a vehicle, all of the platoon's vehicles must be Sent to the Rear (see page 48 of the rulebook) and cannot be Brought Forward again.

You may not Dismount a Passenger-fired .50 cal AA MG or AA MG unless you have the appropriate model to replace the Infantry team that Dismounted it.

Mounting HMG's and LMG's

Up to two HMG or LMG teams carried as Passengers in a half-track can fit their machine-guns to the sides of the vehicle, converting them to side-mounted AA MG's. When the team Dismounts or is Destroyed, its machine-gun goes with it.

INFANTRY

Automatic Rifles

Rifle and Carbine teams do not suffer the usual +1 to hit penalty for firing a ROF 1 weapon while moving or Pinned Down.

When not Pinned Down, Rifle and Carbine teams re-roll all failed rolls to hit in Defensive Fire during assaults.

The Truscott Trot

Infantry and Man-packed Gun teams move 16"/40cm instead of the normal 12"/30cm when moving At the Double.

TANK DESTROYERS

A Self-propelled Anti-tank Platoon, Tank Destroyer Platoon, or Towed Tank Destroyer Platoon is a Reconnaissance platoon.

Each platoon has two components:

- a Security Section consisting of its Jeep, Recon Jeep, and M20 scout car teams and any Carbine teams apart from the Command team of the Tank Destroyer Section, and
- a Tank Destroyer Section of the rest of the platoon.

SECURITY SECTION

If the Platoon Command team of the Security Section is Destroyed, nominate any of the Carbine, Recon Jeep, or M20 scout car teams as the Platoon Command team.

If all of the Carbine, Recon Jeep, and M20 scout car teams of the Security Section are Destroyed, mark the location of the Platoon Command team when it was Destroyed. The Tank Destroyer Section must be placed in the player's next turn as if the Platoon Command team were still where it was when it was Destroyed.

SEEK, STRIKE, AND DESTROY

At the start of the game before deployment you may choose to either:

- deploy your Tank Destroyer Section as normal at the start of the game, leaving the Security Section off table for the whole game, or
- deploy only the Security Section, holding the Tank Destroyer Section off the table ready to be placed later using the Tank Destroyer Section rule.

TANK DESTROYER SECTION

Place the Tank Destroyer Section on the table at the start of any of your turns when Ambushes are revealed so that every team in the Tank Destroyer Section is:

- within Command Distance of a team from the Security Section, and
- In Command when the Security Section is removed.

In addition, each team must not be:

- within 16"/40cm of any enemy team within Line of Sight, unless Concealed by Terrain from it, or
- within 4"/10cm of any enemy team, or
- within 8"/20cm of any enemy Recce team that is in Line of Sight (apart from Recce teams that are Bogged Down, Bailed Out, or moved at the Double).

Once the Tank Destroyer Section is placed, the Security Section is permanently removed from the table. The Tank Destroyer Section can move and fight as normal during the turn it appears. When it takes Platoon Morale Checks, the Tank Destroyer Section ignores any teams from the Security Section that were Destroyed.

AIR OBSERVATION POSTS

COLUMN SECURITY

If a US AOP (see page 139 of the rulebook) did not act as an Observer team in its own turn, then in the opponent's turn, their teams may not be placed from Ambush (see page 266 of the rulebook) within 12"/30cm of the AOP while in Line of Sight of it.

ARTILLERY

EXCELLENT COMMUNICATIONS

Platoon and 2iC Command teams can act as Spotting teams and request artillery fire in the same manner as a Company Command team, suffering the same +1 penalty to Range In.

HIT 'EM WITH EVERYTHING YOU'VE GOT

When a US artillery platoon that has a Staff team fires a bombardment, you may choose to group other artillery platoons into the same Artillery Bombardment before rolling to Range In. Treat all of the weapons as if they were from the same platoon when determining the effect of the Bombardment. Use the Mixed Bombardment rule on page 131 of the rulebook for the combined bombardment.

UNDER COMMAND

Company, 2iC, and any Platoon Command teams do not suffer their normal +1 penalty to Range In when calling for Artillery Bombardments from platoons taken as a Combat or Weapons platoon choice (including Weapons Platoons, Mortar Platoons, and Assault Gun platoons), as well as Cannon Platoons and Chemical Mortar Platoons taken as Support platoon choices.

TIME ON TARGET

If an Artillery Battery with a Staff team Ranges In on their first attempt, they can fire a TOT Bombardment.

Every team hit by a TOT Bombardment must re-roll successful Saves in the first turn of the Bombardment as they scramble for cover and batten down the hatches.

This is for the first turn of the Bombardment only and does not apply to artillery firing an All Guns Repeat! Bombardment.

TURNING TIDE

82ND AND 101ST AIRBORNE DIVISIONS

Gammon Bombs

All teams carrying Gammon Bombs have Tank Assault 3.

Master Sergeant

Parachute Rifle, Airborne Engineer Combat and Airborne Divisional Recon Platoons use the German Mission Tactics special rules (see page 242 of the rulebook).

1ST AND 29TH INFANTRY DIVISIONS ASSAULT COMPANIES

Assault Company

An Assault Company will Always Attacks against any other Infantry Company.

Pioneer Equipment

Rifle teams in Boat Sections count as Pioneer teams when attempting to cross or gap Barbed Wire Entanglements and when assaulting Bunkers.

Clear the Skies

An Assault Company receives a Ground-attack Aircraft Flight on a roll of 6 instead of the usual 5+. However, its Air Support will successfully intercept enemy Ground-Attack Aircraft Flights on a roll of 5+ instead of the usual 6.

2ND AND 5TH RANGER BATTALION

I Shall Never Fail My Comrades

A Ranger Battalion never has to take a Company Morale Check for having more platoons destroyed than still fighting. They take Platoon Morale Checks as normal though. However, if the total number of Ranger Platoons (the combat platoons) Destroyed from the Ranger Battalion exceeds the number still on the table, the Ranger player must roll a die at the start of each turn. On a roll of 3+, the Ranger Battalion fights on heedless of its losses. On any other roll, the game ends. If the Ranger Battalion no longer has a 2iC or Company Command team, then this roll is automatically failed.

No Obstacle Too Tough

Ranger Infantry and Man-packed Gun teams are Mountaineers (see page 61 of the rulebook).

Rangers Lead The Way

Ranger Infantry teams may move At the Double (using the Truscott Trot) through Difficult Going.

RANGERS

2ND, 3RD AND 6TH ARMORED DIVISIONS

SPEARHEAD

Any force from the US 2nd or 3rd Armored Divisions Always Attacks. When they are attacking, the player may make a Spearhead Deployment move (see page 261 of the rulebook) with one platoon that is not a Reconnaissance Platoon.

HEDGEROW CUTTERS

When a tank fitted with a hedgerow cutter attempts to cross a Bocage Hedgerow, it successfully crosses on a roll of 2+ rather than a Skill Test. If it fails, the tank Bogs Down on the hedgerow as normal, but ignores the Belly Up rule (see page 58 of the rulebook).

If it crosses successfully, then any other Fully-tracked Tank teams adjacent to the hedgerow can follow it through on a roll of 2+ instead of a Skill Test as well.

TANK TELEPHONES

Sherman tanks from the US 2nd and 3rd Armored Divisions have Tank Telephones.

If a Tank team with a Tank Telephone and an adjacent Infantry team did not move in the Movement Step and are not Pinned Down, the Infantry team can use the Eyes and Ears rule (see page 195 of the rulebook) to Reveal one Gone to Ground enemy team to that Tank team as if the Infantry team was a Recce team. If other tanks in the platoon fire, they must either have their own Infantry team pointing out the target, or continue to treat the target as Gone to Ground.

TASK FORCE A

TASK FORCE

Task Force A Always Attacks. Combat and Weapons platoons that are not Reconnaissance Platoons may make Spearhead Deployment moves (see page 261 of the rulebook).

NEVER OUT OF TOUCH

Once each turn, you may re-roll one die rolled to receive Reserves for your force.

In a mission using the Scattered Reserves special rule, once per turn you may also re-roll one die rolled to determine where a platoon will arrive from Scattered Reserve.

LIEUTENANT TURNER TURNBULL

Turnbull is a Warrior Command Rifle/MG team and is rated Fearless Veteran. He can replace the Command team in any US Parachute Rifle Platoon for +25 points.

MASTER TACTICIAN

Rifle/MG teams in any platoon commanded by Turnbull may re-roll any failed To Hit rolls when conducting Defensive Fire.

BRIGADIER GENERAL NORMAN 'DUTCH' COTA

Cota is a Warrior Higher Command Carbine team rated as Fearless Veteran.

Cota may join any Assault Company for +25 points.

That's How It's Done

Cota always hits on a roll of 2+ in an assault combat.

Get Off this Beach

Cota starts the game off table. In the Starting Step when you would normally roll for Reserves, roll a die for Cota if he is off the table.

On a roll of 5+, if Cota has not been Destroyed, he is placed adjacent to the Platoon Command team (if present, otherwise any team in the platoon) of any platoon in an Assault Company anywhere on the table.

On any other roll, Cota remains off the table.

Rangers, Lead the Way!

Any platoon led by Cota will pass Motivation tests on a roll of 3+.

I Won't be Able to Do it for You Next Time

In the Starting Step when you would normally roll for Reserves, roll a die if Cota is on the table.

On a roll of 3+, Cota remains where he is.

On any other roll, Cota has finished what he came to do and moves off to another part of the battlefield. Remove him from the table. Next turn roll for him to return again using the Get Off this Beach special rule.

STAFF SERGEANT LAFAYETTE POOL

Pool is a Warrior team. He is rated as Fearless Veteran.

Pool must take over a tank in any Medium Tank Platoon or Medium Tank Platoon (76mm), except the Platoon Command tank, for +25 points. Pool and his tank are no longer part of any platoon.

Hard Charger

If Pool's tank moved at least 6"/15cm towards any enemy team or objective in its previous Movement Step, and is hit in the enemy Shooting Step, roll a die for each hit he is allocated.

- On a score of 5+, Pool's headlong charge puts the enemy off their aim and the hit is ignored.
- Otherwise the hit is resolved as normal.

Eyebrows off a Gnat

Pool's tank may shoot using its Stabiliser without incurring the normal +1 to hit penalty.

82ND AND 101ST AIRBORNE DIVISIONS

GAMMON BOMBS

All teams carrying Gammon Bombs have Tank Assault 3.

MASTER SERGEANT

Parachute Rifle, Airborne Engineer Combat and Airborne Divisional Recon Platoons use the German Mission Tactics special rules (see page 242 of the rulebook).

MAJOR JULIAN COOK

Major Julian Cook is a Warrior Higher Command Rifle team and is rated Fearless Veteran.

Major Cook may join a US Parachute Rifle Company from the 82nd Airborne Division for +50 points.

A DARING PLAN

A force led by Major Julian Cook may include a second Airborne Engineer Platoon as an additional Support Platoon. If you choose to do so, your force will always attack against any other Infantry Company in a mission with the Defensive Battle Special Rule.

AMPHIBIOUS ASSAULT

When crossing rivers using the River Crossing special rules, Parachute Rifle and Airborne Engineer Platoons in a force commanded by Cook that are not Pinned Down may attempt to Get Out of the river during the Assault Step.

- Roll a Motivation Test for each platoon attempting to move out of the river:
- If they pass the Motivation Test, the platoon may Get Out of the river and immediately move or assault up to 4"/10cm.

If they fail the Motivation Test, the platoon cannot move any further this turn and must wait until its next Movement Step to Get Out as normal.

HAIL MARY...

Cook, and any platoon he is currently leading, always passes Motivation Tests on a roll of 2+.

SOFTEN 'EM UP!

Parachute Field Artillery Batteries and Parachute Mortar Platoons in a force commanded by Cook may fire a bombardment using the Soften 'Em Up! special rule. This is a mixed bombardment of high-explosive and smoke shells and must be fired at the beginning of the Shooting Step along with any Smoke Bombardments before any other shooting.

When using the Soften ' Em Up! rule to fire a bombardment with three or more weapons, place a Smoke Screen under the same template used for the bombardment. The bombarding platoon counts as only having two weapons firing and must re-roll successful To Hit rolls. You cannot fire a Soften ' Em Up! bombardment with two or less weapons.

The Smoke Screen is always one template in size and is aligned with the bombardment's template rather than the wind.

As with normal Smoke Bombardments, a bombardment using the Soften 'Em Up special rule cannot use the All Guns Repeat! rule, nor may it use the Time On Target special rule.

DOGS & DEVILS
BREAKOUT AT ANZIO, ITALY, MAY 1944

3RD INFANTRY DIVISION

MEDAL OF HONOR WINNERS

Whenever an Assault or Rifle Platoon from the 3rd Infantry Division is pinned by Defensive Fire in an Assault, roll a die.

On a roll of 3+, one Pioneer Rifle or Rifle team in contact with the enemy does not Fall Back when the rest of the platoon Falls Back. This potential Medal of Honor team may continue the Assault. The Medal of Honor team in contact with the enemy now assaults, hitting on a 2+.

Enemy platoons do not need to make a Motivation Test to Counterattack if only assaulted by a Medal of Honor team.

Regardless of whether or not the Medal of Honor team was hit in the enemy Counterattack, you must take a Motivation Test to Counterattack.

If the platoon passes, the whole platoon may Charge into Contact and continue the assault as normal. The Medal of Honor team has finished its heroics and reverts to a normal team for the rest of the assault.

Otherwise, the Medal of Honor team Breaks Off, returning to the platoon and reverting to a normal team.

SHARPSHOOTER

Any Assault or Rifle Platoon with a Sharpshooter may re-roll any one failed To Hit and any one failed Firepower Test by a Rifle or Pioneer Rifle team in each Shooting Step.

IMPROVISED ARMOUR

See page 118 of the rulebook.

TANK TELEPHONES

See *Turning Tide* special rules on page 8.

STRATEGIC BOMBARDMENT

See *Cassino* special rules on page 13.

FIRE DIRECTION CENTER

See *Cassino* special rules on page 13.

FIRST SPECIAL SERVICE FORCE

I SHALL NEVER FAIL MY COMRADES

A Special Service Force Company ignores all Divisional Support Platoons when determining whether it needs to take a Company Morale Check for being below half strength. Support Platoons never count as being either Destroyed or still on the table for the purposes of a Company Morale Check.

LEAD THE WAY

Special Service Force Infantry teams may move At the Double (using the Truscott Trot) through Difficult Going.

CAUTIOUS MOVEMENT

Special Service Force teams are always Gone to Ground, unless they move At the Double, carry Passengers that are not Recce teams, shoot or assault.

ASSAULT TROOPS

Special Service Force Platoons may re-roll failed Motivation tests to rally from being Pinned Down.

NO OBSTACLE TOO TOUGH

Special Service Force Infantry and Man-packed Gun teams are Mountaineers (see page 61 of the rulebook).

WOODSMEN

Special Service Force Platoons use the German Mission Tactics special rule (see page 242 of the rulebook).

DEATH, DESTRUCTION AND MAYHEM

Special Service Force Carbine, Rifle, and Rifle/MG teams hit on a roll of 2+ in assault combat with a Tank Assault rating of 3.

100TH INFANTRY BAT/442ND REGT COMBAT TEAM 'NISEI'

Sharpshooter

Any Nisei Rifle Platoon with a Sharpshooter may re-roll any one failed roll To Hit and any one failed Firepower Test by a Rifle team in each Shooting Step.

Fighting for Dear Old Uncle Sam

Nisei platoons pass Platoon Morale Checks on a roll of 2+ instead of their normal roll.

A Nisei Company Command team passes Company Morale Checks on a roll of 2+ instead of its normal roll.

Round Up the Huns!

Nisei platoons pass Motivation Tests to rally from being Pinned Down on a roll of 2+ instead of their normal roll.

Go for Broke

Whenever an assaulting Nisei platoon is Pinned Down by Defensive Fire and would normally fall back, immediately roll a Motivation Test.

- If the platoon passes, it is no longer Pinned Down and attempts to carry on with the Assault. The enemy immediately shoots again in Defensive Fire as if this was a new Assault. If the Nisei platoon is not Pinned Down this time it carries on the Assault as normal. If it is Pinned Down a second time by the enemy's Defensive Fire, it remains Pinned Down and falls back as normal.
- Otherwise, the platoon remains Pinned Down and falls back as normal.

91ST CAVALRY RECONNAISSANCE SQUADRON

Rough and Ready Recon

The Rough and Read Recon rule no longer has any effect.

Cavalry Charge

A Cavalry Light Tank Platoon may make a Reconnaissance Deployment move even though it is not a Reconnaissance Platoon.

Mountain Training

All Infantry and Man-packed Gun teams from Dismounted Cavalry Platoons are Mountaineers.

Tank Destroyers

A Cavalry Anti-tank Platoon uses the US Tank Destroyer special rules on page 5.

FIRST SPECIAL SERVICE FORCE

See First Special Service Force special rules from Dogs & Devils on page 11.

FIRE DIRECTION CENTER

Fire Direction Center

Replace any Artillery Staff Team of a Field Artillery Battery (155mm) with a Fire Direction Center team for +50 points.

An FDC team has all the capabilities of a Staff Team plus the Strategic Bombardment special rule.

Strategic Bombardment

A company that has both Air Support and a Fire Direction Center (FDC) may chose to have the Always Attacks special rule. If it does this, is the attacker, and starts the game with the Fire Direction Center (FDC) deployed on the table, it may conduct a Strategic Bombardment during its first turn of the game.

In place of Roll for Air Support in the Starting Step, select any enemy team on the board and place a Ranged In marker on it.

During the attacker's first turn of the game, at the beginning of their Shooting Step, place a devastating bombardment template, parallel with the attacker's table edge, centred on the Ranged In enemy team and have the defender roll for wind direction. (You may choose to re-roll wind direction).

After wind direction has been determined, the attacker rolls a D6 and moves the template's centre 1"/2.5cm times the D6 roll in the direction of the wind.

Roll one die for every team at least partly under the template. A roll of 2+ is a successful hit. Each team under the template gets its normal save, whether Dug-in or not. The Anti-tank rating is 5 and the Firepower is 5+ of the bombardment.

When the bombardment is completed remove one die from the Air Support pool and replace the template with a 12"/30cm square smoke screen that does not disperse until the end of the attacking player's next Starting Step. All weapons starting the game linked to the FDC team are considered having fired for this turn. They cannot move or shoot.

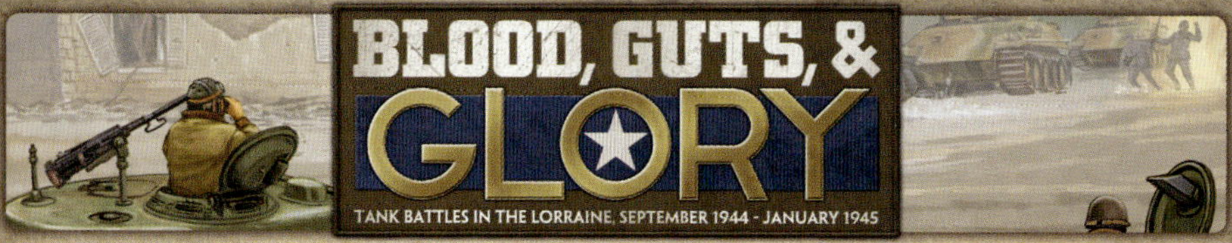

M4 SHERMAN TANK

Detroit's Finest

Tanks that use the Detroit's Finest special rule have a Movement Distance of 14"/35cm on Roads or Cross-country Terrain.

Smooth Ride

An M4A3E8 Easy Eight tank does not suffer the +1 penalty to its score To Hit when using the Stabilizers special rule, provided that it did not move more than 6"/15cm during the Movement Step and it did not move in, enter, or move out of Rough Terrain.

Jumbos Lead the Way

You may allocate hits to an M4A3E2 Jumbo tank as if it had the lowest armour rating, assigning it a hit before the lesser armoured tanks. This rule does not apply to hits from Artillery Bombardments or hits from Aircraft.

Duckbills

At the start of the game a player may elect to fit all of their Sherman tanks (of all variants) with Duckbills. This gives them Wide Tracks (see page 61 of the rulebook), but makes their mobility rating Slow Tank. M4A3E2 Jumbo and M4A3E8 Easy Eight tanks cannot be fitted with Duckbills and ignore the Duckbills special rule.

Tank Telephones

If a Tank team with a Tank Telephone and an adjacent Infantry team did not move in the Movement Step and are not Pinned Down, the Infantry team can use the Eyes and Ears rule (see page 195 of the rulebook) to Reveal one Gone to Ground enemy team to that Tank team as if the Infantry team was a Recce team. If other tanks in the platoon fire, they must either have their own Infantry team pointing out the target, or continue to treat the target as Gone to Ground.

LIEUTENANT GENERAL GEORGE S PATTON JR.

Lieutenant General George S Patton Jr. is a Warrior and Higher Command Transport Team rated Fearless Veteran.

He is mounted in an M20 utility car. You may field Patton for +100 points with the Tank Company on page 22 of *Blood, Guts, & Glory* the Armored Rifle Company on page 26 of *Blood, Guts, & Glory*, or the Tank Destroyer Company on page 32 of *Blood, Guts, & Glory*.

Hold'em by the Nose and Kick'em in the Pants

A force led by Patton has the Always Attacks rule (see page 257 of the rulebook).

Foul Mouth

In any turn in which Patton did not move At the Double, all American platoons with their Platoon Command team in Line Of Sight of Patton pass Motivation Tests on a result of 3+.

The Greatest Soldier

If Patton is Destroyed during the game, American morale at the front and back home suffers a huge loss, despite any other gains. If this happens, the American player loses one Victory Point and their opponent gains one Victory Point at the end of the battle.

A Pint of Sweat

Each turn Patton may re-roll one die rolled to receive Reserves for his company.

Old Blood and Guts

If your force is the Attacker in a mission, each American platoon within Line Of Sight of Patton may make a Spearhead Deployment move (see page 261 of the rulebook) at the beginning of the game.

Lieutenant Colonel Creighton Abrams

Lieutenant Colonel Creighton Abrams is a Warrior Higher Command Tank Team rated Fearless Veteran. He is mounted in one of his "Thunderbolt" tanks as follows:

M4 Sherman "Thunderbolt V"	155 points
M4A3 (76mm) Sherman "Thunderbolt VI"	210 points
M4A3E8 Sherman "Thunderbolt VII"	260 points

You may field Abrams in the Tank Company on page 22 of *Blood, Guts, & Glory* or the Armored Rifle Company on page 26 of *Blood, Guts, & Glory*.

That's the Shortest Way Home!

A force led by Abrams may make Spearhead Deployment moves (see page 261 of the rulebook) with up to two of your platoons at the beginning of the game.

Up Front

Combat and Weapons platoons in a force led by Abrams use the German Mission Tactics special rule (see page 242 of the rulebook).

We're Going In

A platoon led by Abrams may use the German Stormtroopers special rule (see page 241 of the rulebook).

WEB INTELLIGENCE BRIEFINGS

2ND INDIANHEAD DIVISION

Winter Training

While playing under winter conditions, 2nd Infantry Division Infantry and Man-packed Gun teams treat deep snow as Cross-country Terrain instead of Difficult Going, allowing them to move At the Double across deep snow covered terrain.

Tank Telephones

See *Turning Tide* special rules on page 8.

No Obstacle Too Tough

See *Turning Tide* special rules on page 7.

Rangers Lead The Way

See *Turning Tide* special rules on page 7.

Ivory X

You may purchase an Ivory X Staff Team for +25 points. Deploy the Ivory X Staff Team when you would deploy Independent Teams.

Anti-aircraft Artillery Platoons and Towed Tank Destroyer Platoons that have a Command team within 6"/15cm of the Ivory X Staff team can use it as a Staff Team to fire Bombardments.

Anti-aircraft Artillery Platoons and Towed Tank Destroyer Platoons that do not have a Command team within 6"/15cm of the Ivory X Staff Team cannot fire bombardments.

A Bombardment controlled by Ivory X uses all of the American Artillery special rules on page 240 of the rulebook, except for Time On Target.

Since the Ivory X Staff Team is a special team coordinating the task force's artillery fire, we recommend that you use the Forward Direction Center model (UBX23).

Hedgerow Cutters

See *Turning Tide* special rules on page 8.

GLIDER RIFLE COMPANY

Earn Your Jump Boots

Glider Combat and Weapons Platoons may re-roll failed attempts to Dig In.

Gammon Bombs

All teams carrying Gammon Bombs have Tank Assault 3.

Master Sergeant

Parachute Rifle, Airborne Engineer Combat and Airborne Divisional Recon Platoons use the German Mission Tactics special rules (see page 242 of the rulebook).

4TH ARMORED DIVISION

Spearhead

Any force from the US 4th Armored Division Always Attacks. When they are attacking, the player may make a Spearhead Deployment move (see page 261 of the rulebook) with one platoon that is not a Reconnaissance Platoon.

BRIGADIER GENERAL JAMES GAVIN

Brigadier General James Gavin is a Warrior Higher Command Rifle team (Automatic rifle). He is rated Fearless Veteran.

Brigadier General James Gavin may join a US Parachute Rifle Company (from the 82nd Airborne Division found in *Hell's Highway*) or a Glider Rifle Company (found on our website: www.FlamesOfWar.com) for +40 points.

ISOLATE THE BATTLEFIELD

Before your opponent rolls for Reserves you may roll one die for each Combat and Weapons Platoon from your company currently held in Reserve.

- On each roll of 6 your opponent rolls one less die when rolling for Reserves this turn.
- On any other roll, the enemy platoon arrives as normal.

COUNT ON IT

Each time a Parachute Rifle Platoon led by Gavin has Moved at the Double, roll a Skill Check for the platoon:

- If successful, the platoon is not considered to have Moved at the Double during the opponent's next turn,
- Otherwise, the platoon Moves at the Double as normal.

MAJOR GENERAL MAXWELL TAYLOR

Major General Maxwell Taylor is a Warrior Higher Command Carbine team rated Fearless Veteran. Major General Maxwell Taylor may join a US Parachute Rifle Company (from the 101st Airborne Division found in *Hell's Highway*) or a Glider Rifle Company (found on our website: www.FlamesOfWar.com) for +35 points.

INDIAN COUNTRY

A force led by Taylor may make an additional Reconnaissance Deployment move (see page 194 of the rulebook) with any non-reconnaissance platoon (even though its not a Reconnaissance Platoon), provided that the Reconnaissance Deployment move takes the whole platoon out of its deployment area.

BRING UP THE GUNS!

When Taylor is leading an Airborne Anti-tank Platoon, it may use the German Stormtrooper special rule (see page 166 of the rulebook).

LIEUTENANT DANIEL INOUYE

Lieutenant Daniel Inouye is a Warrior Command Pioneer Rifle team and replaces the Command team of one of your Nisei Rifle Platoons. He is a rated Fearless Veteran. Lieutenant Daniel Inouye may join a Niesi Rifle Company (found in Cassino) for +15 points.

HEAVY ASSAULT

Inouye may re-roll his failed Skill tests to Hit in Assault Combat.

HIT 'EM HARD

When Inouye joins one of your Combat or Weapons platoons that launches an Assault, the first hit on the platoon does not count towards Pinning Down the platoon.

US ARSENAL

TANK TEAMS

Name *Weapon*	Mobility *Range*	Armour Front *ROF*	 Side *Anti-tank*	 Top *Firepower*	Equipment and Notes
TANKS					
M5A1 Stuart	Light Tank	4	2	1	Co-ax MG, Hull MG, AA MG.
M6 37mm gun	*24"/60cm*	*2*	*7*	*4+*	*Stabiliser.*
T8E1 Turretless Stuart	Light Tank	4	2	1	Hull .50 cal MG, Hull MG, Wide tracks.
M4 or M4A1 Sherman	Standard Tank	6	4	1	Co-ax MG, Hull MG, .50 cal AA MG.
M3 75mm gun	*32"/80cm*	*2*	*10*	*3+*	*Smoke, Stabiliser.*
M4 or M4A1 Sherman (late)	Standard Tank	6	4	1	Co-ax MG, Hull MG, .50 cal AA MG, Tank telephone.
M3 75mm gun	*32"/80cm*	*2*	*10*	*3+*	*Smoke, Stabiliser. (Blood, Guts, & Glory)*
M4 or M4A Sherman dozer	Standard Tank	6	4	1	Co-ax MG, .50 cal AA MG, Bulldozer.
M3 75mm gun	*32"/80cm*	*2*	*10*	*3+*	*Smoke, Stabiliser.*
M4A1 Sherman DD	Standard Tank	6	4	1	Co-ax MG, .50 cal AA MG, DD tank.
M3 75mm gun	*32"/80cm*	*2*	*10*	*3+*	*Smoke, Stabiliser.*
M4A3 Sherman	Standard Tank	6	4	1	Co-ax MG, Hull MG, .50 cal AA MG, Detroits finest, Tank telephone, Tank telephone.
M3 75mm gun	*32"/80cm*	*2*	*10*	*3+*	*Smoke, Stabiliser.*
M4A3 Sherman (late)	Standard Tank	7	4	1	Co-ax MG, Hull MG, .50 cal AA MG, Detroits finest, Protected ammo, Tank telephone.
M3 75mm gun	*32"/80cm*	*2*	*10*	*3+*	*Smoke, Stabiliser.*
M4A1 (76mm) Sherman	Standard Tank	7	4	1	Co-ax MG, Hull MG, .50 cal AA MG, Protected ammo, Tank telephone.
M1 76mm gun	*32"/80cm*	*2*	*12*	*3+*	*Stabiliser.*
M4A1 (76mm) Sherman (late)	Standard Tank	7	4	1	Co-ax MG, Hull MG, .50 cal AA MG, Protected ammo, Tank telephone.
M1 76mm gun (late)	*32"/80cm*	*2*	*13*	*3+*	*Stabiliser.*
M4A3 (76mm) Sherman	Standard Tank	7	4	1	Co-ax MG, Hull MG, .50 cal AA MG, Detroits finest, Protected ammo, Tank telephone.
M1 76mm gun (late)	*32"/80cm*	*2*	*13*	*3+*	*Stabiliser.*
M4A3E8 Easy Eight	Standard Tank	7	4	1	Co-ax MG, Hull MG, .50 cal AA MG, Detroits finest, Protected ammo, Smooth Ride, Tank telephone, Wide tracks.
M1 76mm gun (late)	*32"/80cm*	*2*	*13*	*3+*	*Stabiliser.*
M4A3E2 Jumbo	Standard Tank	12	8	1	Co-ax MG, Hull MG, .50 cal AA MG, Jumbos Lead the way, Tank telephone.
M3 75mm gun	*32"/80cm*	*2*	*10*	*3+*	*Smoke, Stabiliser.*
LIEUTENANT COLONEL CREIGHTON ABRAM'S TANKS					
M4 "Thunderbolt V"	Standard Tank	6	4	1	Co-ax MG, Hull MG, .50 cal AA MG, Hedgerow cutters, Tank telephone.
M3 75mm gun	*32"/80cm*	*2*	*10*	*3+*	*Smoke, Stabiliser.*
M4A3 (76mm) "Thunderbolt VI"	Standard Tank	7	4	1	Co-ax MG, Hull MG, .50 cal AA MG, Detroits finest, Protected ammo, Tank telephone.
M1 76mm gun (late)	*32"/80cm*	*2*	*13*	*3+*	*Stabiliser.*
M4A3E8 "Thunderbolt VII"	Standard Tank	8	5	1	Co-ax MG, Hull MG, .50 cal AA MG, Protected ammo, Smooth Ride, Tank telephone, Wide tracks.
M1 76mm gun (late)	*32"/80cm*	*2*	*13*	*3+*	*Stabiliser.*

Tank Destroyers

M6 37mm GMC	Wheeled	-	-	-	
M3 37mm gun	*24"/60cm*	*3*	*7*	*4+*	
M3 75mm GMC	Half-tracked	1	0	0	
M1897 75mm gun	*32"/80cm*	*2*	*9*	*3+*	*Hull mounted, Smoke.*
M10 3in GMC	Standard Tank	4	2	0	.50 cal AA MG.
M7 3in gun	*32"/80cm*	*2*	*12*	*3+*	*Slow traverse.*
M10 3in GMC (artillery)	Standard Tank	4	2	0	.50 cal AA MG.
M7 3in gun	*32"/80cm*	*2*	*12*	*3+*	*Slow traverse.*
Firing bombardments	*88"/220cm*	*-*	*3*	*6*	*(Dogs & Devils)*
M10 3in GMC (late)	Standard Tank	4	2	0	.50 cal AA MG.
M7 3in gun (late)	*32"/80cm*	*2*	*13*	*3+*	*Slow traverse.*
M18 Hellcat GMC	Light Tank	2	0	0	.50 cal AA MG.
M1 76mm gun	*32"/80cm*	*2*	*12*	*3+*	
M18 Hellcat GMC (late)	Light Tank	2	0	0	.50 cal AA MG.
M1 76mm gun (late)	*32"/80cm*	*2*	*13*	*3+*	
M36 90mm GMC	Standard Tank	4	2	0	.50 cal AA MG, Detroits finest.
M3 90mm gun	*32"/80cm*	*2*	*14*	*3+*	*Slow traverse.*

Assault Guns

M4 81mm MMC	Half-tracked	1	0	0	
M1 81mm mortar	*24"/60cm*	*2*	*2*	*3+*	*Hull mounted, Portee, Minimum range 8"/20cm, Smoke.*
Firing bombardments	*40"/100cm*	*-*	*2*	*6*	*Smoke bombardment.*
T30 75mm HMC	Half-tracked	1	0	0	
M1A1 75mm pack howitzer	*16"/40cm*	*2*	*6*	*3+*	*Hull mounted, Smoke.*
Firing bombardments	*64"/160cm*	*-*	*3*	*6*	
M8 Scott HMC	Light Tank	3	2	0	.50 cal AA MG.
M1A1 75mm howitzer	*16"/40cm*	*2*	*6*	*3+*	*Smoke.*
Firing bombardments	*64"/160cm*	*-*	*3*	*6*	
M3 75mm/81 mm mortar GMC	Half-tracked	1	0	0	.50 cal AA MG.
M1897 75mm gun	*32"/80cm*	*2*	*9*	*3+*	*Hull mounted, Smoke.*
M1 81mm mortar	*24"/60cm*	*2*	*2*	*3+*	*Hull mounted, Minimum range 8"/20cm, Smoke*
Firing bombardments	*40"/100cm*	*-*	*2*	*6*	
M4 (105mm) Sherman	Standard Tank	7	4	1	Co-ax MG, Hull MG, .50 cal AA MG, Protected ammo.
M4 105mm howitzer	*24"/60cm*	*1*	*9*	*2+*	*Breakthrough gun, Slow traverse, Smoke.*
Firing bombardments	*48"/120cm*	*-*	*4*	*4+*	
M4A3 (105mm) Sherman	Standard Tank	7	4	1	Co-ax MG, Hull MG, .50 cal AA MG, Protected ammo, Detroits finest.
M4 105mm howitzer	*24"/60cm*	*1*	*9*	*2+*	*Breakthrough gun, Slow traverse, Smoke.*
Firing bombardments	*48"/120cm*	*-*	*4*	*4+*	

Armoured Artillery

T19 105mm HMC	Half-tracked	1	0	0	Awkward layout.
M2A1 105mm howitzer	*24"/60cm*	*1*	*9*	*2+*	*Hull mounted, Breakthrough gun, Smoke.*
Firing bombardments	*72"/180cm*	*-*	*4*	*4+*	*Smoke bombardment.*
M7 Priest HMC	Standard Tank	1	0	0	.50 cal AA MG.
M2A1 105mm howitzer	*24"/60cm*	*1*	*9*	*2+*	*Hull mounted, Breakthrough gun, Smoke.*
Firing bombardments	*72"/180cm*	*-*	*4*	*4+*	*Smoke bombardment.*
M12 155mm GMC	Standard Tank	0	0	0	Awkward layout.
M1918M1 155mm gun	*24"/60cm*	*1*	*13*	*1+*	*Hull mounted, Bunker buster.*
Firing bombardments	*104"/260cm*	*-*	*5*	*2+*	
M4 or M4A1 Sherman OP	Standard Tank	6	4	1	Co-ax MG, Hull MG, .50 cal AA MG.
M3 75mm gun	*32"/80cm*	*1*	*10*	*3+*	*Smoke.*
M4A3 Sherman OP	Standard Tank	6	4	1	Co-ax MG, Hull MG, .50 cal AA MG, Detroits finest.
M3 75mm gun	*32"/80cm*	*1*	*10*	*3+*	*Smoke.*
M4A3 Sherman (late) OP	Standard Tank	7	4	1	Co-ax MG, Hull MG, .50 cal AA MG, Detroits finest, Protected ammo.
M3 75mm gun	*32"/80cm*	*1*	*10*	*3+*	*Smoke.*

Self-propelled Anti-aircraft

M13 MGMC (Twin .50 cal)	Half-tracked	1	0	0	
M33 twin .50 cal gun	*16"/40cm*	*5*	*4*	*5+*	*Anti-aircraft.*
M16 MGMC (Quad .50 cal)	Half-tracked	1	0	0	
M45 quad .50 cal gun	*16"/40cm*	*6*	*4*	*5+*	*Anti-aircraft.*
M15 CGMC (37mm)	Wheeled	1	0	0	
M15 37mm combination mount	*24"/60cm*	*4*	*5*	*4+*	*Anti-aircraft.*

Reconnaissance

M8 armored car	Wheeled	1	0	0	Co-ax MG, .50 cal AA MG.
M6 37mm gun	*24"/60cm*	*2*	*7*	*4+*	
M20 scout car	Jeep	1	0	0	.50 cal AA MG.
Recon Jeep	Jeep	-	-	-	
With AA MG	*16"/40cm*	*3*	*2*	*6*	
With .50 cal AA MG	*16"/40cm*	*3*	*4*	*5+*	
Armored Recon Jeep	Jeep	0	0	0	.50 cal AA MG, Overloaded.
Airborne Recon Jeep	Jeep	-	-	-	.50 cal AA MG.
Mortar Jeep	Jeep	-	-	-	
M2 60mm mortar	*24"/60cm*	*2*	*1*	*3+*	*Hull mounted, Portee, Minimum range 8"/20cm.*
Firing Bombardments	*32"/80cm*	*-*	*1*	*6*	
M3A1 armoured car	Jeep	1	0	0	.50 cal AA MG, two AA MG.

Vehicle Machine-guns

Vehicle MG	*16"/40cm*	*3*	*2*	*6*	*ROF 1 if other weapons fire.*
.50 cal Vehicle MG	*16"/40cm*	*3*	*4*	*5+*	*ROF 1 if other weapons fire.*

GUN TEAMS

Weapon	Mobility	Range	ROF	Anti-tank	Firepower	Notes
M2 .50 cal MG	Man-packed	16"/40cm	3	4	5+	
M1919 LMG team	Man-packed	16"/40cm	5	2	6	ROF 2 when pinned down or moving.
M1917 HMG team	Man-packed	24"/60cm	6	2	6	ROF 3 when pinned down or moving.
M2 60mm mortar	Man-packed	24"/60cm	2	1	3+	Minimum range 8"/20cm.
Firing bombardments		32"/80cm	-	1	6	
M1 81mm mortar	Man-packed	24"/60cm	2	2	3+	Smoke, Minimum range 8"/20cm.
Firing bombardments		40"/100cm	-	2	6	Smoke bombardment.
4.2in Chemical mortar	Light	48"/120cm	-	3	4+	Smoke bombardment.
M2 .50 cal AA gun	Heavy	16"/40cm	4	4	5+	Anti-aircraft, Turntable.
M2 .50 cal AA gun (artillery)	Heavy	16"/40cm	4	4	5+	Anti-aircraft, Turntable.
Firing bombardments		40"/100cm	-	-	-	*(Indianhead PDF)*
M49 quad .50 cal AA gun	Heavy	16"/40cm	6	4	5+	Anti-aircraft, Turntable.
M49 quad .50 cal AA gun (artillery)	Heavy	16"/40cm	6	4	5+	Anti-aircraft, Turntable.
Firing bombardments		40"/100cm	-	-	-	*(Indianhead PDF)*
M1 Bofors gun	Immobile	24"/60cm	4	6	4+	Anti-aircraft, Turntable.
M1 Bofors gun (artillery)	Immobile	24"/60cm	4	6	4+	Anti-aircraft, Turntable.
Firing bombardments		40"/100cm	-	-	-	*(Indianhead PDF)*
M3 37mm gun	Light	24"/60cm	3	7	4+	Gun shield.
M1 57mm gun	Medium	24"/60cm	3	10	4+	Gun shield, No HE.
M1 57mm gun (late)	Medium	24"/60cm	3	10	4+	Gun shield.
M5 3in gun	Immobile	32"/80cm	2	12	3+	Gun shield.
M5 3in gun (artillery)	Immobile	32"/80cm	2	12	3+	Gun shield.
Firing bombardments		64"/220cm	-	3	6	*(Dogs & Devils)*
M5 3in gun (late)	Immobile	32"/80cm	2	13	3+	Gun shield.
M1 90mm gun	Immobile	40"/100cm	2	13	3+	Heavy anti-aircraft, Turntable.
M1 90mm gun (artillery)	Immobile	40"/100cm	2	13	3+	Heavy Anti-aircraft, Turntable.
Firing bombardments		88"/220cm	-	3	5+	*(Dogs & Devils)*
M1A1 75mm Pack howitzer	Light	16"/40cm	2	6	3+	Smoke.
Firing bombardments		64"/160cm	-	3	6	Smoke bombardment.
M3 105mm light howitzer	Heavy	16"/40cm	1	7	2+	Breakthrough gun, Smoke.
Firing bombardments		56"/140cm	-	4	4+	Smoke bombardment.
M2A1 105mm howitzer	Immobile	24"/60cm	1	9	2+	Breakthrough gun, Gun shield, Smoke.
Firing bombardments		72"/180cm	-	4	4+	Smoke bombardment.
M1 155mm howitzer	Immobile	24"/60cm	1	10	1+	Bunker buster, Gun shield, Smoke.
Firing bombardments		88"/220cm	-	5	2+	Smoke bombardment.

NAVAL GUN FIRE SUPPORT

Weapon	Range	ROF	Anti-tank	Firepower	Notes
Naval Guns	112"/280cm	-	6	1+	

FORTIFICATIONS

Weapon	Mobility	Range	ROF	Anti-tank	Firepower	Notes
HMG Nest	Immobile	24"/60cm	6	2	6	ROF 3 when Pinned Down.

INFANTRY TEAMS

Team	Range	ROF	Anti-tank	Firepower	Notes
Carbine team	8"/20cm	1	1	6	Automatic rifles.
Rifle team	16"/40cm	1	2	6	Automatic rifles.
Rifle/MG team	16"/40cm	2	2	6	
SMG team	4"/10cm	3	1	6	Full ROF when moving.
Bazooka team	8"/20cm	1	10	5+	Tank assault 4.
Flame-thrower team	4"/10cm	2	-	6	Flame-thrower.
Staff team	16"/40cm	1	2	6	Moves as a Heavy Gun team.
Fire Direction Center	16"/40cm	1	2	6	Moves as a Heavy Gun team.

Additional Training and Equipment

Pioneer teams are rated as Tank Assault 3.

Teams with Gammon Bombs are rated as Tank Assault 3.

TRANSPORT TEAMS

Vehicle *Weapon*	Mobility *Range*	Front *ROF*	Side *Anti-tank*	Top *Firepower*	Equipment and Notes
Trucks					
Jeep or Jeep with trailer	Jeep	-	-	-	Optional Passenger-fired AA MG or .50 cal AA MG.
Dodge ¾-ton or GMC 2½-ton truck	Wheeled	-	-	-	
DUKW truck	Wheeled	-	-	-	Optional Passenger-fired .50 cal AA MG, Amphibious.
M5 high-speed tractor	Standard Tank	-	-	-	.50 cal AA MG.
Armoured Personnel Carriers					
M2 or M3 half-track	Half-tracked	1	0	0	Optional Passenger-fired AA MG or .50 cal AA MG.
M2 37mm half-track	Half-tracked	1	0	0	Passenger-fired .50 cal AA MG.
M3 37mm gun	24"/60cm	3	7	4+	Hull mounted.
M20 utility	Jeep	1	0	0	Passenger-fired .50 cal AA MG.
Recovery and Engineer Vehicles					
M31 TRV recovery vehicle	Standard Tank	5	3	0	Recovery vehicle.
M32 TRV recovery vehicle	Standard Tank	6	4	0	.50 cal AA MG, Recovery vehicle.
Turretless M4 Sherman dozer	Standard Tank	6	4	0	Bulldozer.
Bulldozer	Very Slow Tank	-	-	-	Bulldozer.
Pioneer Supply Truck	Wheeled	-	-	-	
Pioneer Supply Handcart	Wagon	-	-	-	

AIRCRAFT

Aircraft	Weapon	To Hit	Anti-tank	Firepower	Notes
P-38 Lightning	Cannon	3+	7	5+	
	Bombs	4+	5	1+	
P-40 Warhawk	MG	2+	6	5+	
	Bombs	4+	5	2+	
P-47 Thunderbolt	MG	2+	6	5+	
	Bombs	4+	5	1+	

Germany

222

213
215

National Rules

DOCTRINE

STORMTROOPERS

Any German platoon with a Platoon Command team may attempt a Stormtroopers move in its Assault Step.

Roll a Skill Test for each platoon.

- If the test is passed, the platoon treats the Assault Step as a Movement Step in which it can move up to another 4"/10cm, regardless of its normal movement distance.
- If the test is failed, the platoon cannot move any further this turn.

Either way, a platoon that attempts to make a Stormtroopers move cannot take any part in an assault in the same turn.

Although it is not the Movement Step, teams making a Stormtroopers move may Mount and Dismount as if it was the Movement Step. Platoons cannot make Stormtroopers moves if they moved At the Double. Bogged Down or Bailed Out vehicles cannot make Stormtroopers moves.

Only Armoured vehicles can make Stormtroopers moves if they are Pinned Down. Other types of teams cannot make Stormtroopers moves if they are Pinned Down. A team that used Eyes and Ears to reveal Gone to Ground enemy troops cannot make a Stormtroopers move. Immobile Gun teams and Gun teams that shot earlier in the turn cannot make a Stormtroopers move. Teams of any type that have fired an Artillery Bombardment cannot make a Stormtroopers move. Trains cannot make a Stormtroopers move.

REICH DIVIDED

Heer, Luftwaffe, and SS are all treated as Allies (see page 70 of the rulebook). Fallschirmjäger are part of the Luftwaffe, so are not Allies to them, but are Allies to the Heer and SS.

KAMPFGRUPPE

At the start of the game before any platoons are Deployed and before any Combat Attachments are made, any company with a 2iC Command team may take up to half of the teams (counting the Platoon Command team) from any Combat or Weapons platoons in the company and place them in a special Kampfgruppe Platoon. The 2iC Command team is the Command team for this platoon.

You must leave at least two Tank teams or three other teams in each of the Combat or Weapons platoons you take teams from. Teams placed in the Kampfgruppe Platoon are no longer part of their original platoons.

If a platoon attaching troops to the Kampfgruppe has Transport teams, you must take the Transport team that normally carries a team into the Kampfgruppe with it. If that Transport team normally carries other teams as well, they must join the Kampfgruppe as well.

The Kampfgruppe Platoon may include any HQ Support Weapons (including Infantry teams). Any remaining HQ Support Weapons must be attached out as normal. A Kampfgruppe made from the teams of a Fortified Platoon is not a Fortified Platoon and may not contain any of its component platoon's Fortifications. The Kampfgruppe Platoon has the lowest Motivation rating and the lowest Skill rating of any of the teams in it.

MISSION TACTICS

If a Platoon Command Infantry team is Destroyed, another team takes over immediately. Remove any other Infantry team in the platoon that is within Command Distance of the Command team and replace it with the original Platoon Command team. If the Platoon Command team was a Warrior team, it is now just a standard Platoon Command team. If a Platoon Command Tank team is Destroyed, another team takes over immediately. Nominate any other Tank team in the platoon that is within Command Distance of the Destroyed Command team to be the new Platoon Command team. If there is no team of an appropriate type within Command Distance, then the Platoon Command team is Destroyed and the platoon is left leaderless.

The original Platoon Command team can still use this rule while a Warrior team is leading the platoon and acting as its Command team, but the rule does not apply to the Warrior team.

ARMOURED INFANTRY

MOUNTED ASSAULT

Armoured Transport teams in a platoon with the Mounted Assault special rule are Mounted Assault Transports.

A Mounted Assault Transport has a dual nature. While empty, it is just an ordinary Transport team, except that they do not have to be Sent to the Rear when empty—they can remain on the table and continue to fight. When carrying Passengers, a Mounted Assault Transport is treated as a Tank team for everything except Platoon Morale Checks. If it is carrying two or more Passenger teams, it also has Tank Escorts.

As a Tank team, a Mounted Assault Transport team carrying Passengers may Charge into Contact and fight in assault combat. As Open-topped Tank teams, the platoon will be forced to Fall Back by five hits or two Destroyed or Bailed Out vehicles in Defensive Fire (see page 154 of the rulebook).

If a Mounted Assault Transport team is Destroyed by Defensive Fire, all surviving Passengers Fall Back 2"/5cm away from the teams they were charging. If the Passengers cannot do this, they are Destroyed. The platoon is not automatically Pinned Down when the Passengers Dismount, and will only Fall Back if it takes five hits as usual. If a Mounted Assault Transport team is Destroyed during the assault, any Passengers are Destroyed with it.

The Passengers do not need to Dismount to Counterattack. They can stay Mounted or Dismount as they wish. If they Dismount to Counterattack, their Transport team will be Sent to the Rear.

If the platoon voluntarily Breaks Off instead of Counterattacking, and the entire platoon is mounted in Mounted Assault Transport teams, they may use the Break Off Through the Enemy rule (see page 166 of the rulebook) as if they were Tank teams. At the end of the Assault Step, Mounted Assault Transport teams that are Bailed Out are Sent to the Rear if they have no Passengers, or automatically Remount (see page 102 of the rulebook) and are no longer Bailed Out if they have Passengers.

TIGER ACES

At the start of the game before deployment, roll a die for each Warrior or platoon with the Tiger Aces special rule (including any Kampfgruppe made entirely of teams from platoons with the Tiger Aces special rule), then look up their Tiger Ace Skill on the table below.

If the Tiger Ace Skill is specifically rolled for the Platoon Command team or a Warrior team, it only applies to that team unless it is Schnell, Clever Hans, or For the Fatherland, in which case any platoon it is part of also receives the benefit.

SCHNELL!

The platoon may make Stormtroopers moves on a roll of 2+.

CLEVER HANS!

Tanks in this platoon re-roll failed Bogging Checks to cross Rough Terrain and Skill Tests to free a Bogged Down tank. If they have Wide Tracks (see page 61 of the rulebook), they roll to Free themselves (see page 44 of the rulebook) after re-rolling the Bogging Check.

FOR THE FATHERLAND!

The platoon passes Motivation Tests on a roll of 2+.

EVERY SHOT COUNTS!

Tiger tanks in this platoon re-roll any failed roll to hit when they shoot.

RAPID FIRE!

The 8.8cm tank guns of the Tiger tanks in this platoon have ROF 3.

TOP ACE!

Roll again to determine the platoon's Tiger Ace Skill. If you roll 6 again, you may choose the platoon's skill.

In addition, Roll a third time to generate an extra skill specifically for the Platoon Command tank. This gives the platoon commander two Tiger Ace Skills (the one for the platoon as a whole, and their own one specific to them). If you roll a 6 or the same number as the platoon's Tiger Ace Skill, you may choose the platoon commander's extra skill.

TANKS

Schürzen

When a tank that is protected by Schürzen is hit by a weapon with a Firepower of 5+ or 6 on the Side armour by shooting and fails its Armour Save, roll a special 4+ Schürzen save:

- If the save is successful the Schürzen protects the tank from the side shot.
- If the save is not successful the shot penetrated the side armour as normal.

Begleit Escorts

Assault guns with Tank Escorts use all the normal Tank Escort rules. In addition, an assault gun with Tank Escorts may fire its Hull MG at the same time as its hull-mounted main gun. This does not prevent the Tank Escorts from shooting as well.

ARTILLERY

Armoured Rocket Launchers

An Artillery platoon equipped with Armoured Rocket Launcher teams that fired may take a Skill Test in the Assault Step instead of making a Stormtroopers move or taking part in an Assault:

- If they pass the Skill Test, remove the Smoke Trail markers from all Rocket Launcher teams (see page 138 of the rulebook).
- Otherwise, they are too slow in getting away, leaving them vulnerable to counterbattery artillery fire.

Stuka zu Fuss

Each Stuka zu Fuss rocket launcher can only fire one Artillery Bombardment in a game, but counts as six weapons firing when it does so.

You may take up to six attempts to Range In with each Stuka zu Fuss rocket launcher. Each failed attempt reduces the number of weapons firing by one. If you fail all six attempts to Range In, the rockets have all been fired, but missed, and no Bombardment is possible.

If multiple Stuka zu Fuss rocket launchers from the same platoon fire, place a separate Template and roll to Range In separately with each as if they were separate artillery batteries. Complete each Stuka zu Fuss rocket launcher's Bombardment before beginning the next

The half-track is still available to transport its passengers after the Stuka zu Fuss fires.

352. INFANTERIEDIVISION

Fortified Defence

A Festungskompanie or Füsilierkompanie will Always Defend.

A Festungskompnie may purchase fortifications (page 38 of *Earth & Steel*) as part of their force making them a Fortified Company.

Goliath Demolition Carriers

Remove the Goliath remote-control demolition carrier from a team equipped with one if the team moves.

Everything Must Fire On the Beaches

Use the Armoured Reserves rule on page 269 of the rulebook.

21. PANZERDIVISION

Rommel's Zirkus

See the Improvised Vehicle Recovery rules on page 45 of the rulebook.

PANZER LEHR DIVISION

Radio-control Tank Platoon - Königstiger

You may field your Radio-control Tank Platoon with a single Königstiger.

The Königstiger tanks of the Radio-control Tank Platoon are so unreliable that they must take a Skill Test every time they attempt to move.

- If they pass the Skill Test, they can move as normal.
- If they fail the Skill Test, their transmission fails. The Königstiger tank becomes Reluctant and cannot move for the remainder of the game.

A Radio-control Tank Platoon equipped with Königstiger tanks does not have Tiger Ace skills and is not equipped with Borgward BIV demolition carriers.

2. AND 12. SS-PANZERDIVISION

Panzer Ace

Before the game begins, roll a die for the Company Command team of each SS-Panzerkompanie. Look up the Tiger Ace Skills table on page 244 of the rulebook to find the skill possessed by this veteran. If the roll is a 6, simply choose the skill for the commander rather than rolling further. Where the skill refers to Tiger tanks or 8.8cm tank guns, treat it as referring to the commander's own tank and its main gun. The effects of the Tiger Ace Skills only apply to the Company Command team.

2. 'DAS REICH' SS-PANZERDIVISION

INSTANT READINESS

Any platoon entirely made up of Infantry teams, with or without Transport teams, from 2. 'Das Reich' SS-Panzerdivision that is not Pinned Down at the start of the Assault Step, that becomes Pinned Down in the Assault Step, can take a Motivation Test.

- If the test is passed, remove the Pinned Down marker.
- Otherwise, the platoon remains Pinned Down.

If the platoon is Pinned Down by Defensive Fire, they still fall back and the assault is over.

12. 'HITLERJUGEND' SS-PANZERDIVISION

ONE-WAY TRIP TO HEAVEN

When any infantry platoon from 12. 'Hitlerjugend' SS-Panzerdivision fails its Motivation Test to Counterattack during an assault, it has the option to either Break Off as normal or continue the assault.

If the platoon elects to continue the assault, they fight on as if they had passed the Motivation Test. However, if they fail any further Motivation Tests to Counterattack in this assault, then the platoon is immediately Destroyed as if they had failed a Platoon Morale Check.

MAJOR HANS VON LUCK

Major Hans von Luck is a Warrior Higher Command SMG team with a Kübelwagen for transport. He is rated as Confident Veteran.

Von Luck may join a Gepanzerte Panzergrenadierkompanie or Panzergrenadierkompanie from 21. Panzerdivision for +50 points.

Von Luck may be mounted in a Panzer IV H tank instead of being an SMG team for an additional +50 points. If he is mounted in a tank, you may not field one of your Panzer Platoons with more than four Panzer IV H tanks since he has the fifth one.

GET MOVING

Major von Luck, and any platoon he leads, may re-roll failed Skill Tests to make Stormtroopers moves.

BETTER LATE THAN NEVER

Major von Luck always starts the game off table in Reserve. The German player may re-roll one die each turn when rolling for Reserves. When the first unit arrives from Reserve, von Luck arrives with that unit.

If there are no reserves in the mission, von Luck is still in reserve and rolls to arrive as per the Reserves rule with the exception of re-rolling as above. He will then arrive anywhere on the German table edge when successful.

RECONNAISSANCE EXPERT

Unless he has joined a platoon that is not a Reconnaissance Platoon, von Luck may Disengage as if he were a Recce team.

HAUPTMANN KARL-HEINZ NOAK

Hauptmann Karl-Heinz Noak is a Warrior Higher Command Tank team. He may join any company in 654. Schwere Panzerjägerabteilung. Hauptmann Noak is rated Confident Veteran.

Noak's Command Panther

Hauptmann Karl-Heinz Noak is purchased with his Panther A tank for 250 points.

Knight's Cross with Oak Leaves

Hauptmann Karl-Heinz Noak may:

- Move and Shoot with full ROF, and
- have Tank teams in any platoon he joins re-roll failed Skill Tests.

OBERSTURMFÜHRER MICHAEL WITTMANN

Obersturmführer Michael Wittmann is a Warrior Company Command Tank team rated as Fearless Veteran. Wittmann can join any Schwere SS-Panzerkompanie and replaces the Company Command Tiger I E tank team for +100 points. Wittmann takes over the Company Command Tiger tank as his own tank. As Wittmann is a Warrior, he does not roll for Tiger Ace skills.

Every Shot Counts!

Wittmann's Tiger tank may re-roll any failed roll To Hit when it shoots.

Hero of the Fatherland

If Wittmann is Destroyed during a game the morale of the entire German people takes a heavy blow, despite any other gains, and the German player loses one Victory Point and their opponent gains one Victory Point at the end of the battle.

Ambush!

Wittmann may be deployed using the Ambush special rule on page 266 of the rulebook in addition to any other platoons that would normally be deployed in Ambush. Wittmann may deploy in Ambush even in missions that do not normally use the Ambush special rule.

For the Commander!

Wittmann and any platoon led by him always pass Motivation Tests on a roll of 2+.

Grind Them Under

Any team that assaults Wittmann must re-roll the successful Skill Test to hit for any hit allocated to his tank in an assault combat. If they pass this test they get past the tracks and the tank must make an Armour Save as normal. If the re-roll failed, than the assaulting troops have not hit the tank and the hit is ignored.

Rapid Fire!

The 8.8cm tank gun of Wittmann's Tiger tank has ROF 3 when stationary and ROF 2 when moving.

OBERSCHARFÜHRER ERNST BARKMANN

Oberscharführer Barkmann is a Warrior Command Panther A Tank team rated as Fearless Veteran.

Oberscharführer Barkmann may replace an SS-Panzer Platoon with Panther tanks Command Panther A for +60 points in an SS-Panzerkompanie (page 150 of *Earth & Steel*).

Alternatively, he may join a Gepanzerte SS-Panzergrenadier-kompanie (page 154 of *Earth & Steel*), SS-Panzergrenadierkompanie (page 160 of *Earth & Steel*) or SS-Panzerpionierkompanie (page 164 of *Earth & Steel*) for +275 points in place of an SS-Panzer Platoon support choice.

ACE

Barkmann always re-rolls failed To Hit rolls with his Panther's 7.5cm KwK42 gun.

STALKER

If Barkmann is shot at he is counted as Gone to Ground if he is Concealed and did not move in his own turn, even if he did shoot in his turn.

BARKMANN'S WORKSHOP

At the same time as Independent and Warrior teams are deployed, place a Workshop in your deployment area. The Workshop must be in German hands for Barkmann's Panther tank to be repaired. If the enemy take the Workshop (in the same manner they would take an Objective, see page 258 of the rulebook), they capture it and the Workshop is removed from the game.

If Barkmann is Destroyed while mounted in a Panther tank, place his Panther tank adjacent to the Workshop (see above) instead of remaining where it was destroyed.

His platoon may appoint a new Platoon Command team using the Mission tactics special rule as usual when they lose their Platoon Command team and continue to operate without him.

In each Starting Step when you would roll for Reserves to arrive, if there are no enemy teams within 4"/10cm of Barkmann's Panther tank or the Workshop, roll a die.

- On a roll of 5+, his vehicle is repaired and is ready for battle. It is no longer Destroyed. Remove the Workshop when Barkmann is repaired. Barkmann and his repaired Panther A tank are no longer part of his SS-Panzer Platoon.
- Otherwise, the repairs are still progressing.

If the Workshop has been removed or captured, and Barkmann is Destroyed, leave his model one the Table as a Destroyed Armoured Vehicle (see page 103 of the rulebook).

As a Warrior, Barkmann will normally attempt to take over another tank if he is Destroyed. You may choose not to do this and have him Destroyed in his own tank to take advantage of the above rule if you prefer.

78. STURMDIVISION

PAKFRONT!

You may choose your Sturm Anti-tank gun platoon or one of your Sturm Heavy Anti-tank gun platoons to form a Pakfront. Do not deploy the platoon when you would normally. Instead, it is deployed at the same time as Independent Teams.

SPERRVERBAND

TIME'S UP

The Sperrverband HQ, and all Sperr and Sperr Pionier Platoons from a Sperrverband are considered Fearless when taking Motivation Tests to launch an Assault or Counterattack a platoon that contains Armoured Tank teams.

PANZERKNACKER BADGE

All Pioneer teams in a Sperr Pionier Platoon have Tank Assault 5.

ERSATZ PIONIERKOMPANIE

PANZERKNACKER BADGE

All Pioneer teams in a Sperr Pionier Platoon have Tank Assault 5.

SICHERUNGSKOMPANIE

SECURITY TROOPS

Sicherungs Platoons may only be placed in Reserve if no other types of platoon are available to fill the required number of platoons to be held in Reserve.

FELDGENDARMERIE

Feldgendarmerie (Field Police) platoons must make Combat Attachments to Sicherungs Platoons, Gepanzerte Panzergrenadier Platoons, Panzergrenadier Platoons, Gepanzerte Panzerpionier Platoons, Panzerpionier Platoons, Sicherungs Infantry Gun Platoons, Sicherungs Anti-tank Gun Platoons, Motorised Artillery Batteries or Heavy Motorised Artillery Batteries with all their sections.

If a German platoon containing a Feldgendarmerie team that is In Command fails its Motivation Test to rally from being Pinned Down, to Counterattack in an assault, or to pass a Platoon Morale Check (but not to launch an assault against tank or any other Motivation Test), you may immediately Destroy any other Infantry or Gun team from the platoon within Command Distance of the Feldgendarmerie team and re-roll the Motivation Test.

If you roll a 1 for the new Motivation Test, however, the soldiers are pushed past the point of breaking and revolt against the Feldgendarmerie! In this case the Feldgendarmerie team is Destroyed as well.

If a Company Commander joins a platoon with a Feldgendarmerie team attached, the Company Commander overrules the Feldgendarmerie capability. Thus the unit may only use the Company Commander's ability to reroll motivation.

As with any unit, a German platoon Pinned Down by Defensive Fire fails to assault its target. It must wait until its next turn to rally as normal.

ARMOURED TRAIN RULES

See pages 205 to 212 of the rulebook.

TOTENKOPF

Victory at Any Cost

During an assault, any Combat platoon from the 3. SS-Panzerdivision Totenkopf SS-Panzerkampfgruppe or SS-Panzergrenadierkampfgruppe may Destroy one Infantry or Tank team from that platoon that is currently participating in that assault and either re-roll a Motivation Test to Counterattack and apply the new result to the platoon or re-roll a Roll to Hit.

The Destroyed team must be In Command when removed and may not be a Warrior or Independent team. You may do this multiple times in an assault as long as there are teams available to Destroy.

Panzer Ace

If your SS-Panzerkampfgruppe is from 3. SS-Panzerdivision Totenkopf. roll a die before the game begins for the Company Command Tank team. Look up the Tiger Ace Skills table on page 244 of the rulebook to find the acquired skill.

If the roll is a 6, simply choose the skill for the commander rather than rolling further. Where the skill refers to Tiger tanks or 8.8cm tank guns, treat it as referring to the commander's own tank and its main gun.

WIKING

Political Assassins

When Infantry teams from a Gepanzerte SS-Panzergrenadier Platoon or a SS-Panzergrenadier Platoon in a SS-Panzerkampfgruppe or SS-Panzergrenadierkampf gruppe from the 5. SS-Panzerdivision Wiking shoot at a platoon that contains a Soviet Komissar team, you may make a second roll for each hit scored on the platoon.

If you roll a 5 or 6, you can choose to mark the Komissar team as a priority target similar to the Gun Tank rule in the rulebook.

Otherwise, the hits are allocated as normal.

When used, this rule overrides all other rules about hit allocation except those regarding valid targets.

Danish and Flemish Regiments

A SS-Panzerkampfgruppe or SS-Panzergrenadierkampfgruppe from the 5. SS-Panzerdivision Wiking must be made up of either Danish or Flemish troops.

Master Plan (Danish)

When firing artillery bombardments using Motorised, Motorised Heavy, Armoured or Armoured Heavy SS-Artillery Batteries in support of the Danish Panzergrenadier Regiment, Roll to Hit as if the artillery are Veterans.

Following Orders (Flemish)

Platoons from the Flemish Regiment may make a Stormtroopers move while Pinned Down and may move towards the enemy while making this move.

NEDERLAND

Instant Readiness

Any SS-Freiwilligen-Panzergrenadier Platoon, SS-Freiwilligen-Aufklärungs Platoon, or SS-Panzergrenadier Pioneer Platoon from a Nederland SS-Freiwilligen-Panzergrenadierkompanie, that is not Pinned Down before the Assault Step, that becomes Pinned Down during the Assault Step, can take a Motivation Test.

- If the test is passed, remove the Pinned Down marker.
- Otherwise, the platoon remains Pinned Down.

If the platoon is Pinned Down by Defensive Fire, they still fall back and the assault is over.

Following Orders

Platoons from a Nederland SS-Freiwilligen-Panzergrenadierkompanie may make a Stormtroopers move while Pinned Down and may move towards the enemy while making this move.

NORDLAND

Master Plan

When firing artillery bombardments using Motorised SS-Artillery Batteries or Motorised Heavy SS-Artillery Batteries in support of a Nordland SS-Freiwilligen-Panzergrenadierkompanie, Roll to Hit as if the artillery are Veterans.

To the Bitter End

When any SS-Freiwilligen-Panzergrenadier Platoon, SS-Freiwilligen-Aufklärungs Platoon, or SS-Panzergrenadier Pioneer Platoon from a Nordland SS-Freiwilligen-Panzergrenadierkompanie fails its Motivation Test to Counterattack during an assault, it has the option to either Break Off as normal or continue the assault.

If the platoon elects to continue the assault, they fight on as if they had passed the Motivation Test. However, if they fail any further Motivation Tests to Counterattack in this assault, then the platoon is immediately Destroyed as if they had failed a Platoon Morale Check.

FELDHERRNHALLE

Street Brawlers

Feldherrnhalle Company Command, 2iC Command and Platoon Command Infantry teams are Street Brawlers.

If there are no enemy Tank teams or Bunkers within 2"/5cm of a Street Brawler, the Street Brawler hits on a roll of 2+ in Assaults.

FLORIAN GEYER AND MARIA THERESIA

Local Knowledge

A SS-Kavalerieschwadron may deploy a German or Hungarian platoon entirely made up of Infantry or Gun teams, or a mix of both, in Immediate Ambush (see page 267 of the rulebook) in addition to any Ambush deployment that may be allowed by the Mission played. The Immediate Ambushing platoon must be taken from the platoons that are to be deployed on table during deployment.

VETERAN TANK-HUNTER PLATOON

Shoot and Scoot

Veteran Tank-hunter Platoons pass Skill Tests to make a Stormtroopers move on a roll of 2+.

Manoeuvre and Fire

Veteran Tank-hunter Platoons may prepare alternate firing positions. At the start of the game choose a single piece of Concealing Area Terrain or Linear Obstacle that provides Concealment. If a Veteran Tank-hunter Platoon is placed from Ambush choose the terrain feature at the start of the Starting Step in the turn the ambushing platoon is placed. The selected terrain feature must be at least partially within your deployment area.

As long as the whole platoon remains Concealed within or behind the selected terrain feature, they may shoot using their full ROF when they move, as if they had not moved. If the platoon leaves the selected terrain feature, or is no longer Concealed from enemy teams other than aircraft, they lose the benefit of the Manoeuvre and Fire special rule for the remainder of the game.

ARMOURED ROCKET LAUNCHER BATTERY

DEVASTATING BOMBARDMENT

See the Rocket Launchers rules on page 138 of the rulebook.

SS-VIELFACHWERFER BATTERY

DEVASTATING BOMBARDMENT

See the Rocket Launchers rules on page 138 of the rulebook.

STURMTIGERS

ROCKET ASSAULT HOWITZER

A Sturmtiger may fire as an Artillery Battery. Each Sturmtiger fires as a separate Artillery Battery. Each Sturmtiger only makes one attempt to Range In on its target. Once Ranged In, roll one die To Hit for each target under the template, as normal for an Artillery Bombardment. The score needed To Hit is always 5+ with no modifiers for number of guns firing and no re-rolls. Armoured vehicles make an Armour Save based on their Top armour. All other teams are Destroyed on successful hits.

BUNKERFEUER

If a Sturmtiger shoots at a Bunker or Building, it Ranges In on a roll of 5+. If it successfully Ranges In on a Bunker, the Bunker is automatically Destroyed. If it successfully Ranges In on a Building, every team in the Building is automatically Destroyed.

MOTORCYCLE RECONNAISSANCE

See the Motorcycle Reconnaissance rules on pages 196 to 197 of the rulebook.

FELDWEBEL DIETRICH UTHOFF

Feldwebel Dietrich Uthoff is a Warrior Panzerfaust Assault Rifle team. He is rated as Fearless Veteran.

Uthoff may join a Sturmkompanie from the 78. Sturmdivision for +50 points.

AMBUSH!

Uthoff may be deployed using the Ambush special rule in the rulebook in addition to any other platoons that would normally be deployed in Ambush.

While defending, Uthoff may deploy in Ambush even in missions that do not normally use the Ambush special rule.

TANK HUNTER

Feldwebel Uthoff may move and shoot with his Panzerfaust anti-tank launcher.

FAUST EXPERT

Feldwebel Uthoff may re-roll failed To Hit rolls with his Panzerfaust anti-tank launcher.

OBERSTLEUTNANT DR FRANZ BÄKE

Franz Bäke is a Warrior Higher Command Tank team rated as Fearless Veteran. He may join a Kampfgruppe Bäke company mounted in a Panther tank replacing the Company Command team for +60 points.

ENERGETIC COMMANDER

Bäke and any platoon he is attached to may make Stormtroopers move on a roll of 2+.

MASTER TACTICIAN

If Bäke is present on the battlefield one Bäke Panther Platoon or Bäke Schwere Panzer Platoon may make a Reconnaissance Deployment move after deployment is completed but before the first turn begins.

EVERY SHOT COUNTS

Bäke's Panther tank may re-roll any failed To Hit rolls when it shoots.

FELDWEBEL LUDWIG WINDGRUBER

Windgruber replaces the 2iC in a Sperrverband or Grenadierkompanie for +30 points. He is a Warrior and a Pioneer Panzerfaust SMG team rated Confident Veteran.

GREEN HELL (GRÜN HÖLLE)

All teams from a Sperr, Sperr Pionier, or Grenadier Platoon joined by, or in a Kampfgruppe formed by, Feldwebel Windgruber hit on 2+ in assaults.

OLD HARE (ALTER HÄSE)

Feldwebel Windgruber may launch an Assault from 6"/15cm away from enemy teams. He moves up to 6"/15cm when Charging into Contact or Counterattacking. Any platoon he has joined still move 4"/10cm in assaults, often meaning that they don't get into the fight until the platoon Counterattacks.

STURMBANNFÜHRER FRITZ BIERMEYER

Sturmbannführer Fritz Biermeyer is a Warrior Command Tank team who leads a Panzer Platoon in a Totenkopf SS-Panzerkampfgruppe or SS-Panzergrenadierkampfgrppe. You must purchase him and his platoon together. Biermeyer and his platoon are rated Fearless Veteran.

TOWARDS THE ENEMY

In the Assault Step, a platoon led by Biermeyer must assault if they can. If they cannot assault they must attempt to make a Stormtroopers move as far as possible towards the closest enemy team or closest Objective.

BIERMEYER'S 3 SS-PANZER PLATOON

Biermeyer and his platoon cannot be joined by Warrior or Independent teams. The platoon is organized as a standard Panzer Platoon as on page 109 of *Grey Wolf* with Biermeyer as the Platoon Command Tank for the following point costs.

5 Panzer IV H	325 points
4 Panzer IV H	260 points
3 Panzer IV H	195 points

RECKLESS

Sturmbannführer Biermeyer and his platoon are hit on a 3+ as if they were Trained.

NEW RECRUITS

If Biermeyer is Destroyed then the platoon becomes Confident Trained for the remainder of the game.

GENERAL DER PANZERTRUPPEN DIETRICH VON SAUCKEN

General der Panzertruppen von Saucken is mounted in a Sd Kfz 250 half-track. He is a Warrior Higher Command SMG team rated as Fearless Veteran. Von Saucken may join a German company for +65 points.

COUNTERATTACK EXPERT

All platoons within Command Distance of von Saucken may re-roll any failed Skill Tests to make a Stormtroopers move.

PANZERS MARCH

Each turn von Saucken may re-roll one die rolled to receive Reserves for his company.

PANZER KANONEN

You may replace any or all Platoon Command teams from any SS-Panzer Platoons in the Wiking SS-Division with a Panzer Kanone for +65 points per platoon. A Panzer Kanone is a Warrior with the following abilities:

PANZER KANONE

At the start of the game roll a die for each Panzer Kanone to determine the medal he has been awarded. That Panzer Kanone possesses the skills shown for that medal in the table below.

KNIGHT'S CROSS

A Panzer Kanone with a Knight's Cross may:

Have Tank teams in their platoon re-roll failed Skill Tests.

KNIGHT'S CROSS WITH OAK LEAVES

A Panzer Kanone with a Knight's Cross with Oak Leaves may:

Move and Shoot with full ROF, and

Have Tank teams in their platoon re-roll failed Skill Tests.

KNIGHT'S CROSS WITH SWORDS

A Panzer Kanone with a Knight's Cross with Swords may:

Re-roll any failed roll To Hit when they shoot with their main gun,

Move and Shoot with full ROF, and

Have Tank teams in their platoon re-roll failed Skill Tests.

KNIGHT'S CROSS WITH DIAMONDS

A Panzer Kanone with a Knight's Cross with Diamonds may:

Re-roll any failed roll To Hit when they shoot with their main gun,

Move and Shoot with full ROF,

Only be destroyed on an enemy roll of 5+ rather than a roll of 4+ when using the Warrior Tank Team Casualties rule in the rulebook, and

Have Tank teams in their platoon re-roll failed Skill Tests.

OBERSTLEUTNANT FREIHERR VON DER HEYDTE

Oberstleutnant Freiherr von der Heydte is a Warrior Higher Command SMG team rated as Fearless Veteran.

Von der Heydte may join a Fallschirmjägerkompanie from the 6. Fallschirmjäger Regiment for +70 points. Add up to four Panzerfaust Trap teams for +25 points per team.

YOU ARE RELIEVED!

Platoons from the SS and the Heer that are under Von der Heydte's command are not considered Allied platoons.

RALLY ONCE MORE

Once per turn Von der Heydte may attempt to reform a team from his force. If an Infantry or Gun team within 6"/15cm of Von der Heydte is Destroyed, roll a die:

- If the result is 5+, Von der Heydte rallies the survivors of the team and it continues to fight on unharmed.
- Otherwise, the soldiers are too badly wounded and the team is Destroyed as normal.

NEVER SURRENDER

Von der Heydte, and any platoon he is currently leading, always passes Motivation Tests on a roll of 2+.

PANZERFAUST TRAP TEAMS

Panzerfaust Trap teams are Infantry teams and are rated Fearless Veteran. They are armed with Panzerfaust anti-tank launchers with Range 4"/10cm, ROF 1, Anti-tank 12, Firepower 5+, and Tank Assault 6. A Panzerfaust Trap team cannot shoot in the Shooting Step if it moved in the Movement Step.

WELL HIDDEN

So long as a Panzerfaust Trap team remains where it was deployed it always counts as Concealed and in Bulletproof Cover even when it shoots. If a Panzerfaust Trap team moves for any reason, it no longer benefits from Well Hidden.

FANATICAL FAUST-MEN

Panzerfaust Trap teams are Independent Teams. However, they may never Join a platoon or other Independent Teams. Also, unlike normal Independent Teams, they may Charge into Contact and Counterattack in assaults.

PANZERFAUST AMBUSH

Panzerfaust Trap teams are never deployed at the start of the game. Instead at the start of any of their Movement Steps, the owning player may place any of their Panzerfaust Trap teams in any terrain feature in their own deployment area or No Man's Land such that the Panzerfaust Trap team is not within 2"/5cm of any enemy team.

Panzerfaust Trap teams may not deploy within 8"/20cm and in Line of Sight of a Recce team unless the Recce team moved At the Double their last turn.

If there are any friendly teams within 4"/10cm, Panzerfaust Trap teams may not fire or Launch an Assault and must immediately Break Off when assaulted.

SS-KAMPFGRUPPE GRAEBNER

Reinforced Truck

If a Reinforced Truck is hit, Infantry teams carried as passengers survive their Passenger Save on a roll of 3+.

KAMPFGRUPPE VON TETTAU

Ostfront Veterans

Rifle teams from SS-Unteroffizier Schule 'Arnheim' are Tank Assault 4.

Tettau's Legion

Unlike the SS-Unteroffizier Schule 'Arnhem' and the Hermann Göring Training and Replacement Regiment, platoons from a Sicherheit company do not operate as separate platoons, rather as a whole company. Wherever the rules talk about a platoon, read that as a Sicherheit Company. When you are fielding a Sicherheit Battalion, read rules that talk about a company, as referring to your Sicherheit Battalion.

SS-KAMPFGRUPPE SPINDLER

Kampfgruppe Marsch!

Once each turn, you may re-roll one die rolled to receive reserves for your force. In a Mission using the Scattered Reserves special rule, once per turn you may re-roll one die rolled to determine where a platoon will arrive from Reserve.

No Fear of Tanks

An SS-Panzer Crew Platoon does not need to pass a Motivation Test to assault an enemy platoon that has Armoured Tank teams as it normally would under the Tank Terror rule on page 143 of the rulebook. An SS-Panzer Crew Platoon that suffers no hits in an assault against an enemy platoon that has Armoured Tank teams, does not need to pass a Motivation Test to Counterattack as it normally would under the Must Test if Assaulted by Tanks rule on page 162 of the rulebook.

SS-HAUPTSTURMFÜHRER VIKTOR GRAEBNER

Graebner is a Warrior Company Command Tank team mounted in either a Humber scout car or a Humber IV armoured car. Graebner replaces the Company Command team of SS-Kampfgruppe Graebner and is rated Fearless Veteran.

Graebner may join SS-Kampfgruppe Graebner for +65 points.

Graebner's Gamble

When attacking in a mission with a force led by Graebner, your force makes a dash for the objective. Immediately before Reconnaissance Deployment movement, roll a Skill Test for each of your Combat and Weapons platoons on the table. If successful, the platoon may use the Spearhead Deployment rule on page 261 of the rulebook. Graebner automatically passes this Skill Test.

Knight's Cross

Graebner and any platoon that he is currently leading re-roll failed Skill Tests.

Schnell!

When Graebner and any Combat or Weapons platoon he is currently leading makes a Stormtroopers move, it may move up to 6"/15cm.

HERMANN GÖRING FALLSCHIRMPANZERDIVISION

ELITE TROOPS

Herman Göring Platoons fight to the last and may re-roll any failed Platoon Morale Checks. Your Company Command team may re-roll Company Morale Checks.

MOTORISED MOVEMENT

The Motorised Movement rule no longer has any effect.

EAST FRONT EXPERIENCE

A HG Heavy Anti-aircraft Gun Platoon has the Horse Artillery special rule on page 118 of the rulebook.

INFANTRY DEFENCE

When you deploy your HG Light Anti-aircraft Gun Platoon, you may designate them as infantry support. The 2cm FlaK38 guns designated as infantry support are not anti-aircraft guns, instead they are Light Guns that may fire self-defense anti-aircraft. This means they can be concealed when Dug In like normal light guns.

STACHELDRAHT

A Panzer III M tank equipped with concertina wire on their rear decks may perform a Stacheldraht manoeuvre.

Once per game you may complete a Stacheldraht manoeuvre during the Assault Phase. First, roll for a Stormtroopers move as normal, and if successful, a Panzer III M tank with concertina wire can deploy two Barbed Wire Entanglements instead of moving. Next, place one entanglement on both sides of the Panzer III M extending 8"/20cm perpendicular with the tank's facing. (There is a gap between the two entanglements the width of the deploying tank).

While carrying concertina wire, the Tank's main gun is considered a hull-mounted weapon. The Tank also acts as a Pioneer Supply Vehicle for any platoons within Command Distance attempting to Dig In.

If any terrain or teams block the deployment of the wire then the Stacheldraht manoeuvre cannot be attempted.

OBERST FERDINAND GOSEWISCH

Oberst Gosewisch with his Regimental Command Headquarters is a Warrior Higher Command Staff team and replaces the Staff team of a Motorised Artillery Battery for +40 points (page 72 of *Dog & Devils*). Oberst Gosewisch is rated Confident Veteran.

When Oberst Gosewisch and his team are deployed as part of the Motorised Artillery Battery, they act as a Staff Team and the battery may re-roll Skill Tests for Digging In. Successful Skill Tests for Digging In for the battery places all Gun teams in the battery in Gun Pits. If the battery starts the game in Prepared Positions then all teams in the battery receive Overhead Camouflage and all Gun teams are in Gun Pits at the start of the game.

EXCELLENT COMMUNICATIONS

Command teams in Combat or Weapons Platoons and the 2iC Command team in a Grenadierkompanie containing Oberst Gosewisch can act as spotting teams and request artillery fire as if a Company Command team.

OVERHEAD CAMOUFLAGE

The following artillery batteries and anti-aircraft platoons in a Grenadierkompanie within 6"/15cm of Gosewisch's Regimental Command Headquarters may re-roll failed Skill Tests to Dig In and if successful also receive Overhead Camouflage.

- Motorised Artillery Batteries
- Rocket Launcher Batteries
- HG Heavy Anti-aircraft Gun Platoons
- Grenadier Light Anti-aircraft Gun Platoons

Teams with Overhead Camouflage are Concealed. In addition they are also Concealed to Aircraft.

CASSINO
ITALY: JANUARY - MAY 1944

1. FALLSCHIRMJÄGERDIVISION

Never Surrender

Combat and Weapons platoons of a Fallschirmjäger-kompanie pass Platoon Morale Checks on a roll of 2+ instead of their normal roll.

A Fallschirmjäger Company Command team passes Company Morale Checks on a roll of 2+ instead of its normal roll.

90. PANZERGRENADIERDIVISION

Motorised Movement

The Motorised Movement rule no longer has any effect.

RESERVE PANTHER PLATOON

Von Mackensen's Reserve

In missions that have Reserves of any type (Reserves, Scattered Reserves, Delayed Reserves, etc.), the Reserve Panther Platoon must be held in Reserve and is always the last platoon to arrive from Reserve.

In missions that do not normally have any form of Reserves, the Reserve Panther Platoon starts the game in Reserve.

PANZER BRIGADE

Always Attack

A FHH Panzerkompanie (page 56 of *Blood, Guts, and Glory*), FHH Panzersturmkompanie (page 60 of *Blood, Guts, and Glory*), FHH Panzergrenadierkompanie (page 66 of *Blood, Guts, and Glory*), Panzerkompanie (page 72 of *Blood, Guts, and Glory*), or Panzergrenadierkompanie (page 74 of *Blood, Guts, and Glory*) uses the Always Attacks rule on page 257 of the rulebook.

Tailor-made Battlegroups

A FHH Panzerkompanie (page 56 of *Blood, Guts, and Glory*), FHH Panzersturmkompanie (page 60 of *Blood, Guts, and Glory*), FHH Panzergrenadierkompanie (page 66 of *Blood, Guts, and Glory*), Panzerkompanie (page 72 of *Blood, Guts, and Glory*), or Panzergrenadierkompanie (page 74 of *Blood, Guts, and Glory*) cannot use the Kampfgruppe special rule on page 242 of the rulebook.

Web Intelligence Briefings

ESTONIAN-SS

The Line Must Not Break!

If defending in a Mission with the Defensive Battle rule, an Estonian player may hold one platoon in Immediate Ambush (see page 267 of the rulebook). This is in addition to any ambush platoons allowed by the mission played.

For Estonia!

If defending in a Mission with the Defensive Battle rule an Estonian Company Command team may re-roll failed Company Morale Checks.

Devastating Bombardment

See the Rocket Launchers rules on page 138 of the rulebook.

17-SS GÖTZ VON BERLICHINGEN

One-way Trip To Heaven

When any SS-Panzergrenadier Platoon, SS-Scout Platoon, SS-Pionier Platoon, or SS-Aufklärungs Platoon from 17. SS-Panzergrenadier-division 'Götz von Berlichingen' fails its Motivation Test to Counterattack during an assault, it has the option to either Break Off as normal or continue the assault.

If the platoon elects to continue the assault, they fight on as if they had passed the Motivation Test. However, if they fail any further Motivation Tests to Counterattack in this assault, then the platoon is immediately Destroyed as if they had failed a Platoon Morale Check.

Instant Readiness

A Gepanzerte SS-Panzergrenadier Platoon from 2. 'Das Reich' SS-Panzerdivision, that is not Pinned Down before the Assault Step, that becomes Pinned Down during the Assault Step, can take a Motivation Test.

- If the test is passed, remove the Pinned Down marker.
- Otherwise, the platoon remains Pinned Down.

If the platoon is Pinned Down by Defensive Fire, they still fall back and the assault is over.

KAMPFGRUPPE KÄSTNER

Korsun Pocket Breakout Force

A Kampfgruppe Kästner Grenadierkompanie Always Attacks in any mission.

Wiking SS Special Rules

See the Following Orders and Political Assassins rule from *Grey Wolf* on page 32.

Night Attack

In any mission in which a Kampfgruppe Kästner Grenadierkompanie is attacking it may opt for a Night Attack.

If they do so, the game starts in darkness and uses the Night Fighting rules (on page 272 of the rulebook) until morning breaks using the Dawn rules (page 273 of the rulebook).

ELEFANTS — 653. SCHWERE PANZERJÄGERABTEILUNG

One-way Trip To Heaven

When any Panzergrenadier Platoon or Panzerpionier Platoon from 9. SS-Panzerdivision 'Hohenstaufen' fails its Motivation Test to Counterattack during an assault, it has the option to either Break Off as normal or continue the assault.

If the platoon elects to continue the assault, they fight on as if they had passed the Motivation Test. However, if they fail any further Motivation Tests to Counterattack in this assault, then the platoon is immediately Destroyed as if they had failed a Platoon Morale Check.

Devastating Bombardment

See the Rocket Launchers rules on page 138 of the rulebook.

5. GEBIRGSDIVISION

Mountaineers

Gebirgsjäger Infantry teams, Man-packed Gun teams, and Gun teams carried by Pack Mule teams, are all Mountaineers (see page 61 of the rulebook).

Pack Mules

Gun teams carried by Pack Mule teams are Mountaineers. Gun teams carried by Pack Mule teams move as Man-packed. Pack Mule teams don't have to be modelled.

MAJOR ROBERT KÄSTNER

Robert Kästner is a Warrior Higher Command Panzerfaust SMG team rated as Confident Veteran. He may join a Kampfgruppe Kästner Grenadierkompanie for +50 points. He can be mounted in a Kubelwagen for +5 points.

Keep Moving Forward

Kästner and any platoon he has joined may re-roll any failed bogging check to cross Difficult Terrain.

Kästner and any platoon he has joined may make a Stormtroopers move on a roll of 2+.

Expert Fieldcraft

Kästner and any teams in a platoon he has joined are Gone to Ground, unless they move At the Double, carry Passengers that are not Recce teams, shoot, or assault.

Lead from the Front

If Kästner is present on the battlefield he and any one platoon he has joined may make Spearhead Deployment move (see page 261 of the rulebook).

HAUPTMANN SIEGFRIED JAMROWSKI

Hauptmann Siegfried Jamrowski is a Warrior Higher Command SMG team rated Fearless Veteran. Hauptmann Siegfried Jamrowski may join a Fallschirmjägerkompanie (found in *Cassino*) for +70 points.

I'M ALREADY HERE!

As a Warrior team Jamrowski can take and hold objectives (see page 258 of the rulebook).

UNSHAKEABLE

Your Combat and Weapons Platoons fight to the last and may re-roll any failed Platoon Morale Checks. Jamrowski may re-roll Company Morale Checks.

STICK IT TO 'EM!

In the Starting Step, when Jamrowski joins a platoon that is Pinned Down, that platoon automatically rallies from being Pinned Down.

German Arsenal

TANK TEAMS

Name *Weapon*	Mobility *Range*	Armour Front *ROF*	Side *Anti-tank*	Top *Firepower*	Equipment and Notes
TANKS					
Panzer III J (early)	Standard Tank	5	3	1	Co-ax MG, Hull MG, Protected ammo.
5cm KwK38 gun	*24"/60cm*	*3*	*7*	*4+*	
Panzer III L	Standard Tank	6	3	1	Co-ax MG, Hull MG, Protected ammo.
5cm KwK39 gun	*24"/60cm*	*3*	*9*	*4+*	
Panzer III M	Standard Tank	6	3	1	Co-ax MG, Hull MG, Protected ammo, Schürzen.
5cm KwK39 gun	*24"/60cm*	*3*	*9*	*4+*	
Panzer III N	Standard Tank	6	3	1	Co-ax MG, Hull MG, Protected ammo, Schürzen.
7.5cm KwK37 gun	*24"/60cm*	*2*	*9*	*3+*	*Smoke.*
Firing Bombardments	*48"/120cm*	*-*	*2*	*6*	
Panzer IV G (late)	Standard Tank	6	3	1	Co-ax MG, Hull MG, Protected ammo.
7.5cm KwK40 gun	*32"/80cm*	*2*	*11*	*3+*	
Panzer IV H	Standard Tank	6	3	1	Co-ax MG, Hull MG, Protected ammo, Schürzen.
7.5cm KwK40 gun	*32"/80cm*	*2*	*11*	*3+*	
Panzer IV J	Standard Tank	6	3	1	Co-ax MG, Hull MG, Protected ammo, Schürzen.
7.5cm KwK40 gun	*32"/80cm*	*2*	*11*	*3+*	*Slow traverse.*
Panzer IV/70 (A)	Slow Tank	8	3	1	Hull MG, Overloaded, Schürzen.
7.5cm PaK42 gun	*32"/80cm*	*2*	*14*	*3+*	*Hull mounted.*
Panzer IV/70 (V)	Slow Tank	9	3	1	Hull MG, Overloaded, Schürzen.
7.5cm PaK42 gun	*32"/80cm*	*2*	*14*	*3+*	*Hull mounted.*
Panther D	Standard Tank	10	5	1	Co-ax MG, Hull MG, Wide tracks, Unreliable.
7.5cm KwK42 gun	*32"/80cm*	*2*	*14*	*3+*	
Panther A or G	Standard Tank	10	5	1	Co-ax MG, Hull MG, Wide tracks.
7.5cm KwK42 gun	*32"/80cm*	*2*	*14*	*3+*	
Tiger I E	Slow Tank	9	8	2	Co-ax MG, Hull MG, Protected ammo, Wide tracks.
8.8cm KwK36 gun	*40"/100cm*	*2*	*13*	*3+*	*Slow traverse.*

Königstiger (Porsche)	Slow Tank	14	8	2	Co-ax MG, Hull MG, Overloaded, Unreliable.
8.8cm KwK43 gun	*40"/100cm*	*2*	*16*	*3+*	*Slow traverse.*
Königstiger (Henschel)	Slow Tank	15	8	2	Co-ax MG, Hull MG, Overloaded.
8.8cm KwK43 gun	*40"/100cm*	*2*	*16*	*3+*	*Slow traverse.*
Flammpanzer III	Standard Tank	6	3	1	Co-ax MG, Hull MG.
1.4cm Flammenwerfer	*4"/10cm*	*4*	*-*	*5+*	*Flame-thrower.*
Tiger (P) (late)	Slow Tank	13	8	2	Co-ax MG, Hull MG, Protected ammo, Wide tracks, Unreliable.
8.8cm KwK36 gun	*40"/100cm*	*2*	*13*	*3+*	*Slow traverse.*
Bergepanther (Panzer IV turret)	Standard Tank	8	4	1	2 Hull MG, Wide tracks
7.5cm KwK40 gun	*32"/80cm*	*2*	*11*	*3+*	*Hull mounted.*
CAPTURED TANKS					
Panzer 18-R 730(f) (FT-18)	Very Slow Tank	1	0	1	Unreliable.
3.7cm KwK18 (f) gun	*16"/40cm*	*2*	*4*	*4+*	*One-man turret.*
Panzer 35-R 731(f) (R-35)	Slow Tank	3	3	1	Co-ax MG.
3.7cm KwK18 (f) gun	*16"/40cm*	*2*	*4*	*4+*	*One-man turret.*
Panzer 38-H 735(f) (H-39)	Standard Tank	3	3	1	Co-ax MG.
3.7cm KwK38 (f) gun	*24"/60cm*	*2*	*5*	*4+*	*One-man turret.*
Panzer 35-S 739 (f) (S-35)	Standard Tank	4	3	1	Co-ax MG.
4.7cm KwK35 (f) gun	*24"/60cm*	*2*	*6*	*4+*	*One-man turret.*
Panzer B-2 740(f) (Char B-1)	Slow Tank	6	5	2	Co-ax MG, Unreliable.
4.7cm KwK35 (f) gun	*24"/60cm*	*2*	*6*	*4+*	*One-man turret.*
7.5cm KwK29 (f) gun	*16"/40cm*	*2*	*6*	*3+*	*Awkward layout, Hull mounted.*
Flamepanzer B-2 (f) (Char B-1)	Slow Tank	6	5	1	Co-ax MG, Unreliable.
4.7cm KwK35 (f) gun	*24"/60cm*	*2*	*6*	*4+*	*One-man turret.*
1.4cm Flammenwerfer	*4"/10cm*	*3*	*-*	*5+*	*Flame-thrower, Hull mounted.*
M5A1 Stuart	Light Tank	4	2	1	Co-ax MG, Hull MG, AAMG.
M6 37mm gun	*24"/60cm*	*2*	*7*	*4+*	
M4, M4A1 or Sherman III	Standard Tank	6	4	1	Co-ax MG, Hull MG.
M3 75mm gun	*32"/80cm*	*2*	*10*	*3+*	*Smoke.*
Firefly VC	Standard Tank	6	4	1	Co-ax MG.
OQF 17 pdr gun	*32"/80cm*	*2*	*13*	*3+*	*No HE.*
Panzer M14/41	Standard Tank	3	2	1	Co-ax MG, Twin hull MG.
4.7cm KwK 47/32 gun	*24"/60cm*	*2*	*7*	*4+*	
T-70 obr 1943 (captured)	Standard Tank	4	2	1	Co-ax MG, Limited vision, Wide tracks, Unreliable.
4.5cm KwK(r) gun	*24"/60cm*	*1*	*7*	*4+*	
T-34 obr 1942 (captured)	Standard Tank	6	5	1	Co-ax MG, Hull MG, Fast tank, Wide tracks, Limited vision, Unreliable.
7.62cm KwK(r) gun	*32"/80cm*	*2*	*9*	*3+*	
T-34/85 obr 1943 (captured)	Standard Tank	7	5	1	Co-ax MG, Hull MG, Unreliable.
8.5cm KwK(r) gun	*32"/80cm*	*2*	*12*	*3+*	
M4 (M4A2 Sherman) (captured)	Standard Tank	6	4	1	Co-ax MG, Hull MG, Unreliable.
7.5cm KwK(a) gun	*32"/80cm*	*2*	*10*	*3+*	
ASSAULT GUNS					
39H OP	Standard Tank	2	2	0	AA MG, Unreliable.
7.5cm (Sf) 39H	Standard Tank	2	2	0	AA MG, Unreliable.
7.5cm PaK40 gun	*32"/80cm*	*2*	*12*	*3+*	*Hull mounted.*
10.5cm (Sf) 39H	Standard Tank	2	2	0	AA MG, Unreliable.
10.5cm leFH16 gun	*32"/80cm*	*2*	*10*	*2+*	*Hull mounted, Breakthrough gun.*
Firing bombardments	*72"/180cm*	*-*	*4*	*4+*	*Smoke bombardment.*
StuG M42 75/18 (Semovente 75/18)	Standard Tank	4	2	1	
7.5cm StuK 75/18 gun	*24"/60cm*	*2*	*9*	*3+*	*Hull mounted.*
StuG M42 75/34 (Semovente 75/34)	Standard Tank	4	2	1	AA MG.
75/34 gun	*32"/80cm*	*2*	*10*	*3+*	*Hull mounted.*
StuG M43 105/25 (Semovente 105/25)	Standard Tank	6	2	1	AA MG, Overloaded.
105/25 gun	*32"/80cm*	*2*	*10*	*2+*	*Hull mounted, Breakthrough gun.*
StuG D or E	Standard Tank	5	3	1	AA MG, Protected ammo.
7.5cm StuK36 gun	*24"/60cm*	*2*	*9*	*3+*	*Hull mounted.*
StuG G	Standard Tank	7	3	1	Hull MG, Protected ammo, Schürzen.
7.5cm StuK40 gun	*32"/80cm*	*2*	*11*	*3+*	*Hull mounted.*
StuH42	Standard Tank	7	3	1	Hull MG, Protected ammo, Schürzen.
10.5cm StuH42 gun	*32"/80cm*	*2*	*10*	*2+*	*Hull mounted, Breakthrough gun, Smoke.*

StuG IV	Standard Tank	7	3	1	Hull MG, Protected ammo, Schürzen.
7.5cm StuK40 gun	*32"/80cm*	*2*	*11*	*3+*	*Hull mounted.*
Brummbär	Slow Tank	9	5	1	Hull MG, Overloaded, Schürzen.
15cm StuH43 gun	*16"/40cm*	*1*	*13*	*1+*	*Hull mounted, Bunker buster.*
Sturmtiger	Slow Tank	12	8	2	Hull MG, Overloaded.
Firing bombardments	*48"/120cm*	*-*	*6*	*1+*	*Rocket assault howitzer, Bunkerfeuer.*

Infantry Guns (SP)

Sd Kfz 250/7 (8cm)	Half-tracked	1	0	0	AA MG.
8cm GW34 mortar	*24"/60cm*	*2*	*2*	*3+*	*Hull mounted, Portee, Smoke, Minimum range 8"/20cm.*
Firing Bombardments	*40"/100cm*	*-*	*2*	*6*	*Smoke bombardment.*
Sd Kfz 250/8 (7.5cm)	Half-tracked	1	0	0	AA MG.
7.5cm KwK37 gun	*24"/60cm*	*2*	*9*	*3+*	*Hull mounted.*
Sd Kfz 251/2 (8cm)	Half-tracked	1	0	0	AA MG.
8cm GW34 mortar	*24"/60cm*	*2*	*2*	*3+*	*Hull mounted, Portee, Smoke, Minimum range 8"/20cm.*
Firing Bombardments	*40"/100cm*	*-*	*2*	*6*	*Smoke bombardment.*
Sd Kfz 251/9 (7.5cm)	Half-tracked	1	0	0	AA MG.
7.5cm KwK37 gun	*24"/60cm*	*2*	*9*	*3+*	*Hull mounted.*
Sd Kfz 251/9 D (7.5cm) (late)	Half-tracked	1	0	0	Co-ax MG.
7.5cm KwK37 gun	*24"/60cm*	*2*	*9*	*3+*	*Hull mounted. (Blood, Guts, & Glory)*
Sd Kfz 251/16 (Flamm)	Half-tracked	1	0	0	Hull MG.
Two 1.4cm Flammenwerfer	*4"/10cm*	*3 (each)*	*-*	*6*	*Side mounted, Flame-thrower.*
Grille (15cm sIG) H	Standard Tank	2	1	0	AA MG.
15cm sIG33 gun	*16"/40cm*	*1*	*13*	*1+*	*Hull mounted, Bunker buster.*
Firing bombardments	*56"/140cm*	*-*	*4*	*2+*	
Grille (15cm sIG) K	Standard Tank	0	0	0	AA MG.
15cm sIG33 gun	*16"/40cm*	*1*	*13*	*1+*	*Hull mounted, Bunker buster.*
Firing bombardments	*56"/140cm*	*-*	*4*	*2+*	
S307(f) Reihenwerfer	Half-tracked	1	0	0	AA MG, Unreliable.
8cm Reihenwerfer	*40"/100cm*	*-*	*2*	*6*	*Rocket launcher, Smoke bombardment.*
Sd Kfz 251/1 D (Stuka) half-track	Half-tracked	1	0	0	Hull MG.
28cm sW40 Rocket Launcher	*40"/100cm*	*-*	*3*	*1+*	*Hull mounted, Stuka zu Fuss. (Blood, Guts, & Glory)*

Tank-hunters

S307(f) (PaK40)	Half-tracked	1	0	0	AA MG, Unreliable.
7.5cm PaK40 gun	*32"/80cm*	*2*	*12*	*3+*	*Hull mounted.*
Marder I	Slow Tank	0	0	0	AA MG, Overloaded.
7.5cm PaK40 gun	*32"/80cm*	*2*	*12*	*3+*	*Hull mounted.*
Marder II	Standard Tank	1	0	0	AA MG.
7.5cm PaK40 gun	*32"/80cm*	*2*	*12*	*3+*	*Hull mounted.*
Marder III	Standard Tank	1	0	0	Hull MG.
7.62cm PaK36(r) gun	*32"/80cm*	*2*	*12*	*3+*	*Hull mounted.*
Marder III H	Standard Tank	1	0	0	Hull MG.
7.5cm PaK40 gun	*32"/80cm*	*2*	*12*	*3+*	*Hull mounted.*
Marder III M	Standard Tank	0	0	0	AA MG.
7.5cm PaK40 gun	*32"/80cm*	*2*	*12*	*3+*	*Hull mounted.*
PaK40 auf RSO	Slow Tank	0	0	0	
7.5cm PaK40 gun	*32"/80cm*	*2*	*12*	*3+*	*Hull mounted.*
Hetzer	Standard Tank	7	2	1	Hull MG, Overloaded
7.5cm PaK39 gun	*32"/80cm*	*2*	*11*	*3+*	*Hull mounted.*
Jagdpanzer IV	Standard Tank	7	3	1	Hull MG, Protected ammo, Schürzen.
7.5cm StuK40 gun	*32"/80cm*	*2*	*11*	*3+*	*Hull mounted.*
Hornisse	Standard Tank	1	1	0	AA MG, Protected ammo.
8.8cm PaK43 gun	*40"/100cm*	*2*	*16*	*3+*	*Hull mounted.*
Jagdpanther (early)	Standard Tank	10	5	1	Hull MG, Unreliable.
8.8cm PaK43 gun	*40"/100cm*	*2*	*16*	*3+*	*Hull mounted. (Earth & Steel)*
Jagdpanther	Standard Tank	10	5	1	Hull MG.
8.8cm PaK43 gun	*40"/100cm*	*2*	*16*	*3+*	*Hull mounted.*
Elefant	Slow Tank	15	8	2	Hull MG, Overloaded, Unreliable.
8.8cm PaK43 gun	*40"/100cm*	*2*	*16*	*3+*	*Hull mounted.*

Artillery (SP)

Wespe	Standard Tank	1	1	0	AA MG, Protected ammo.
10.5cm leFH18M howitzer	*24"/60cm*	*1*	*10*	*2+*	*Hull mounted, Breakthrough gun, Smoke.*
Firing bombardments	*72"/180cm*	-	*4*	*4+*	*Smoke bombardment.*
Hummel	Standard Tank	1	1	0	AA MG, Protected ammo.
15cm sFH18 howitzer	*24"/60cm*	*1*	*13*	*1+*	*Hull mounted, Bunker buster, Smoke.*
Firing bombardments	*80"/200cm*	-	*5*	*2+*	*Smoke bombardment.*
10.5cm (Sf) Lorraine Schlepper	Slow Tank	0	0	0	AA MG, Overloaded, Unreliable.
10.5cm leFH18 howitzer	*24"/60cm*	*1*	*10*	*2+*	*Hull mounted, Breakthrough gun, Smoke.*
Firing bombardments	*72"/180cm*	-	*4*	*4+*	*Smoke bombardment.*
15cm (Sf) Lorraine Schlepper	Slow Tank	0	0	0	AA MG, Overloaded, Unreliable.
15cm sFH13 howitzer	*16"/40cm*	*1*	*13*	*1+*	*Hull mounted, Bunker buster, Smoke.*
Firing bombardments	*64"/160cm*	-	*5*	*2+*	*Smoke bombardment.*
Panzer II OP	Standard Tank	3	1	1	Co-ax MG, Protected ammo.
2cm Kwk38 gun	*16"/40cm*	*3*	*5*	*5+*	
Panzer III OP	Standard Tank	5	3	1	Hull MG.
Lorraine Schlepper OP	Slow Tank	0	0	0	AA MG, Overloaded, Unreliable.
Panzerwerfer 42 (Maultier)	Half-tracked	0	0	0	AA MG, Armoured rocket launcher.
15cm RW42 rocket launcher	*64"/160cm*	-	*3*	*4+*	*Rocket launcher, Smoke bombardment.*
S307(f) R-Vielfachwerfer	Half-tracked	1	0	0	AA MG, Armoured rocket launcher, Unreliable.
8cm rocket launcher	*56"/140cm*	-	*2*	*6*	*Rocket launcher.*
Vielfachwerfer auf Maultier	Half-tracked	1	0	0	AA MG, Armoured rocket launcher.
8cm rocket launcher	*56"/140cm*	-	*2*	*6*	*Rocket launcher.*

Anti-aircraft (SP)

Sd Kfz 10/5 (2cm)	Half-tracked	-	-	-	Gun shield
2cm FlaK38 gun	*16"/40cm*	*4*	*5*	*5+*	*Anti-aircraft.*
Armoured Sd Kfz 10/5 (2cm)	Half-tracked	0	0	0	
2cm FlaK38 gun	*16"/40cm*	*4*	*5*	*5+*	*Anti-aircraft.*
Sd Kfz 251/17 (2cm)	Half-tracked	1	0	0	
2cm FlaK38 gun	*16"/40cm*	*4*	*5*	*5+*	*Anti-aircraft. (Blood, Guts, & Glory)*
Sd Kfz 251/21 (Triple 15mm)	Half-tracked	1	0	0	
MG151/15 gun	*16"/40cm*	*6*	*4*	*5+*	*Anti-aircraft. (Blood, Guts, & Glory)*
U304(f) (2cm)	Half-tracked	0	0	0	Unreliable.
2cm FlaK38 gun	*16"/40cm*	*4*	*5*	*5+*	*Anti-aircraft.*
3.7cm FlaK43 on Opel truck	Wheeled	-	-	-	Gun shield
3.7cm FlaK43 gun	*24"/60cm*	*4*	*6*	*4+*	*Anti-aircraft, Portee.*
Sd Kfz 7/1 (Quad 2cm)	Half-tracked	-	-	-	Gun shield
2cm FlaK38 (V) gun	*16"/40cm*	*6*	*5*	*5+*	*Anti-aircraft.*
Armoured Sd Kfz 7/1 (Quad 2cm)	Half-tracked	0	0	0	
2cm FlaK38 (V) gun	*16"/40cm*	*6*	*5*	*5+*	*Anti-aircraft.*
Sd Kfz 7/2 (3.7cm)	Half-tracked	-	-	-	Gun shield
3.7cm FlaK43 gun	*24"/60cm*	*4*	*6*	*4+*	*Anti-aircraft.*
Armoured Sd Kfz 7/2 (3.7cm)	Half-tracked	0	0	0	
3.7cm FlaK43 gun	*24"/60cm*	*4*	*6*	*4+*	*Anti-aircraft.*
Flakpanzer 38(t) (2cm)	Standard Tank	0	0	0	
2cm FlaK38 gun	*16"/40cm*	*4*	*5*	*5+*	*Anti-aircraft.*
Möbelwagen (3.7cm)	Standard Tank	0	0	0	
3.7cm FlaK43 gun	*24"/60cm*	*4*	*6*	*4+*	*Anti-aircraft.*
Wirbelwind (Quad 2cm)	Standard Tank	3	1	0	Hull MG.
2cm FlaK38 (V) gun	*16"/40cm*	*6*	*5*	*5+*	*Anti-aircraft.*
2cm FlaK (V) on a Panzer IV	Standard Tank	0	0	0	
2cm FlaK38 (V) gun	*16"/40cm*	*6*	*5*	*5+*	*Anti-aircraft.*
Flakpanzer T-34(r)	Standard Tank	3	2	0	Hull MG, Fast tank, Wide-tracks.
2cm FlaK38 (V) gun	*16"/40cm*	*6*	*5*	*5+*	*Anti-aircraft.*
Bergepanther (Quad 2cm)	Standard Tank	4	2	0	Hull MG, Wide tracks.
2cm FlaK38 (V) gun	*16"/40cm*	*6*	*5*	*5+*	*Anti-aircraft.*

Reconnaissance

Sd Kfz 221 (MG)	Wheeled	0	0	0	AA MG.
Sd Kfz 221 (2.8cm)	Wheeled	0	0	0	
2.8cm sPzB41 anti-tank rifle	*16"/40cm*	*2*	*7*	*5+*	*Hull mounted, No HE.*
Sd Kfz 222 (2cm)	Wheeled	1	0	0	Co-ax MG.
2cm KwK38 gun	*16"/40cm*	*3*	*5*	*5+*	*Self-defence anti-aircraft.*
Sd Kfz 223 (radio)	Wheeled	1	0	0	AA MG.
Sd Kfz 231 (8-rad)	Jeep	2	0	0	Co-ax MG.
2cm KwK38 gun	*16"/40cm*	*3*	*5*	*5+*	
Sd Kfz 234/2 Puma	Jeep	3	0	0	Co-ax MG
5cm KwK39 gun	*24"/60cm*	*2*	*9*	*4+*	
Sd Kfz 250 (recce)	Half-tracked	1	0	0	Hull MG, AA MG.
Sd Kfz 250/9 (2cm)	Half-tracked	1	0	0	Co-ax MG.
2cm KwK38 gun	*16"/40cm*	*3*	*5*	*5+*	*Self-defence anti-aircraft.*
Panzer II L Luchs	Light Tank	3	1	1	Co-ax MG.
2cm KwK38 gun	*16"/40cm*	*3*	*5*	*5+*	
Panhard 178 (f)	Wheeled	1	1	0	Co-ax MG, Limited vision.
2.5cm KwK(f) gun	*16"/40cm*	*2*	*6*	*5+*	*No HE.*
BA-10M	Wheeled	1	0	0	Co-ax MG, Hull MG, Limited vision, Unreliable.
4.5cm KwK(r) gun	*24"/60cm*	*2*	*7*	*4+*	
M8 Greyhound	Wheeled	1	0	0	Co-ax MG, .50cal AA MG.
M6 37mm gun	*24"/60cm*	*2*	*7*	*4+*	
Humber IV	Wheeled	1	0	0	Co-ax MG.
M6 37mm gun	*24"/60cm*	*2*	*7*	*4+*	
Humber Scout Car	Jeep	1	0	0	AA MG.

Reconnaissance Support

Sd Kfz 233 (7.5cm)	Jeep	2	0	0	AA MG.
7.5cm KwK37 gun	*24"/60cm*	*2*	*9*	*3+*	*Hull mounted.*

Motorcycle Reconnaissance

Motorcycle MG team	Jeep	-	-	-	Motorcycle reconnaissance, Dismount as MG team.
MG	*16"/40cm*	*3*	*2*	*6*	*Hull-mounted, Vehicle MG.*
Motorcycle Panzerfaust SMG team	Jeep	-	-	-	Motorcycle reconnaissance, Dismount as Panzerfaust SMG team.
When firing as SMG	*4"/10cm*	*3*	*1*	*6*	*Hull-mounted, Vehicle MG.*
When firing as Panzerfaust	*4"/10cm*	*1*	*12*	*5+*	*Awkward layout.*
Schwimmagen MG team	Jeep	-	-	-	Amphibious, Motorcycle reconnaissance, Dismount as MG team.
MG	*16"/40cm*	*3*	*2*	*6*	*Hull-mounted, Vehicle MG.*
Schwimmagen Panzerfaust SMG team	Jeep	-	-	-	Amphibious, Motorcycle reconnaissance, Dismount as Panzerfaust SMG team.
When firing as SMG	*4"/10cm*	*3*	*1*	*6*	*Hull-mounted, Vehicle MG.*
When firing as Panzerfaust	*4"/10cm*	*1*	*12*	*5+*	*Awkward layout.*

Armoured Trains

Locomotive	Train	3	3	2	Locomotive.
Artillery Car	Train	3	3	2	Artillery Car, Protected ammo, Four passenger-fired side train MG (two each side).
10.5cm leFH18/40 howitzer	*24"/60cm*	*1*	*10*	*2+*	*Deck turret, Breakthrough gun, Smoke.*
Firing bombardments	*72"/180cm*	*-*	*4*	*4+*	*Smoke bombardment.*
Anti-aircraft Car	Train	3	3	2	Artillery Car, Protected ammo, Four passenger-fired side train MG (two each side).
10.5cm leFH18/40 howitzer	*24"/60cm*	*1*	*10*	*2+*	*Deck turret, Breakthrough gun, Smoke.*
Firing bombardments	*72"/180cm*	*-*	*4*	*4+*	*Smoke bombardment.*
with 2cm FlaK38 (V) gun	*16"/40cm*	*6*	*5*	*5+*	*Turntable, Anti-aircraft.*
with 3.7cm FlaK43 gun	*24"/60cm*	*4*	*6*	*4+*	*Turntable, Anti-aircraft.*
Tank-hunter Car	Train	6	3	1	Artillery car, Co-ax MG, Protected ammo, Schürzen.
7.5cm KwK40 gun	*32"/80cm*	*2*	*11*	*3+*	
Infantry Car	Train	3	3	2	Infantry car, Four passenger-fired train MG (2 each side)
Staff Car	Train	3	3	2	Infantry car, Four passenger-fired train MG (2 each side)
Light Artillery Car	Train	3	3	2	Artillery Car, Six train MG, AA MG turret.
Two 7.5cm FK02/26 (p)	*24"/60cm*	*2*	*8*	*3+*	*Deck turrets.*
Firing bombardments	*72"/180cm*	*-*	*3*	*6*	
with 2cm FlaK38 (V) gun	*16"/40cm*	*6*	*5*	*5+*	*Turntable, Anti-aircraft.*
Heavy Artillery Car	Train	3	3	2	Artillery car, Six train MG, AA MG turret.
7.5cm FK02/26 (p) gun	*24"/60cm*	*2*	*8*	*3+*	*Deck turret.*
Firing bombardments	*72"/180cm*	*-*	*4*	*5+*	
10cm leFH14/19 (p) howitzer	*24"/60cm*	*1*	*8*	*2+*	*Deck turret, Breakthrough gun, Smoke.*
Firing bombardments	*72"/180cm*	*-*	*4*	*5+*	*Smoke bombardment.*
Assault Car	Train	3	3	2	Infantry car, Four train MG, Passengers.
Panzer 38(t) S	Standard Tank	3	1	1	Co-ax MG, Protected ammo.
3.7cm KwK38(t) gun	*24"/60cm*	*2*	*6*	*4+*	

Vehicle Machine-guns

Vehicle MG	*16"/40cm*	*3*	*2*	*6*	*ROF 1 if other weapons fire.*
Train MG	*16"/40cm*	*2*	*2*	*6*	*Retain ROF 2 if other weapons fire.*
AA MG turret	*16"/40cm*	*2*	*2*	*6*	*Self-defence anti-aircraft.*

FORTIFICATIONS

Bunkers and Pillboxes

Weapon	Mobility	Range	ROF	Anti-tank	Notes	
8.8cm PaK43 Pillbox	Immobile	40"/100cm	2	16	3+	
7.5cm PaK40 Pillbox	Immobile	32"/80cm	2	12	3+	
7.5cm PaK40 Nest	Immobile	32"/80cm	2	12	3+	
7.5cm (or 7.62cm) FK Pillbox	Immobile	24"/60cm	2	10	3+	
5cm PaK38 Pillbox	Immobile	24"/60cm	3	9	4+	
5cm KwK Nest	Immobile	24"/60cm	3	7	4+	
2cm FlaK38 Nest	Immobile	16"/40cm	4	5	5+	Anti-aircraft.
Flak Nest	Immobile	16"/40cm	6	5	5+	ROF 3 when pinned down, Anti-aircraft
Quad 2cm FlaK Nest	Immobile	16"/40cm	6	5	5+	Anti-aircraft.
HMG Pillbox	Immobile	24"/60cm	6	2	6	ROF 3 when pinned down.
HMG Nest	Immobile	24"/60cm	6	2	6	ROF 3 when pinned down.
28cm Naval Gun Pillbox	Immobile	112"/280cm	-	6	1+	Naval gun Battery.
28cm sWG41 Nest	Immobile	40"/100cm	-	3	1+	Static rocket launcher.

Turrets

			Armour		
Turret ***Weapon***	**Mobility** ***Range***	**Front** ***ROF***	**Side** ***Anti-tank***	**Top** ***Firepower***	**Equipment and Notes**
Panzer I Turret	Immobile	1	1	1	Twin MG.
Panzer II Turret	Immobile	3	1	1	Turret bunker MG.
2cm KwK38 gun	*16"/40cm*	*3*	*5*	*5+*	
R-35 Turret	Immobile	3	3	1	Turret bunker MG.
3.7cm KwK18 (f) gun	*16"/40cm*	*2*	*4*	*4+*	
APX Turret	Immobile	5	4	1	Turret bunker MG.
4.7cm KwK35 (f) gun	*24"/60cm*	*2*	*6*	*4+*	

T-70 Turret	Immobile	4	2	1	Turret bunker MG.
4.5cm KwK38(r) gun	*24"/60cm*	*1*	*7*	*4+*	
T-34/76 Turret	Immobile	6	5	1	Turret bunker MG.
7.62cm KwK34(r) gun	*32"/80cm*	*2*	*9*	*3+*	
Panther Turret	Immobile	10	5	1	Turret bunker MG.
7.5cm KwK42 gun	*32"/80cm*	*2*	*14*	*3+*	

Turret Bunker Machine-guns

Turret Bunker MG	*16"/40cm*	*4*	*2*	*6*	*Cannot shoot if main gun fires.*

GUN TEAMS

Weapon	Mobility	Range	ROF	Anti-tank	Firepower	Notes
MG42 HMG	Man-packed	24"/60cm	6	2	6	ROF 3 when pinned down or moving.
7.5cm LG40 recoilless gun	Man-packed	16"/40cm	2	9	3+	Recoilless.
8.8cm RW43 (Püppchen) launcher	Man-packed	16"/40cm	1	11	5+	
8cm GW42 (Stummelwerfer) mortar	Man-packed	24"/60cm	2	2	3+	Smoke, Minimum range 8"/20cm.
Firing bombardments		32"/80cm	-	2	6	Smoke bombardment.
8cm GW34 mortar	Man-packed	24"/60cm	2	2	3+	Smoke, Minimum range 8"/20cm.
Firing bombardments		40"/100cm	-	2	6	Smoke bombardment.
10.5cm NbW35 mortar	Man-packed	40"/100cm	-	3	4+	Smoke bombardment.
12cm sGW43 mortar	Light	56"/140cm	-	3	3+	
7.5cm leIG18 gun	Light	16"/40cm	2	9	3+	Gun shield, Smoke.
Firing bombardments		48"/120cm	-	3	6	
15cm sIG33 gun	Heavy	16"/40cm	1	13	1+	Bunker buster, Gun shield.
Firing bombardments		56"/140cm	-	4	2+	
2cm FlaK38 gun	Light	16"/40cm	4	5	5+	Anti-aircraft, Gun shield, Turntable.
2cm Flakvierling 38 gun	Immobile	16"/40cm	6	5	5+	Anti-aircraft, Gun shield, Turntable.
3.7cm FlaK43 gun	Immobile	24"/60cm	4	6	4+	Anti-aircraft, Gun shield, Turntable.
3.7cm PaK36 gun	Light	24"/60cm	3	6	4+	Gun shield.
Firing Stielgranate		8"/20cm	1	12	5+	
4.7cm PaK177(i) gun (47/32)	Man-packed	24"/60cm	3	7	4+	
4.7cm PaK183(f) gun (SA-37)	Medium	24"/60cm	3	9	4+	Gun shield, No HE.
5cm PaK38 gun	Medium	24"/60cm	3	9	4+	Gun shield.
7.5cm PaK97/38 gun	Medium	24"/60cm	2	10	3+	Gun shield.
7.5cm PaK40 gun	Medium	32"/80cm	2	12	3+	Gun shield.
7.5cm PaK50 gun	Medium	24"/60cm	2	10	3+	Gun shield.
Firing bombardments		64"/160cm	-	3	6	
7.62cm PaK36(r) gun	Heavy	32"/80cm	2	11	3+	Gun shield.
Captured OQF 6 pdr gun	Medium	24"/60cm	3	10	4+	Gun shield.
8.8cm FlaK36 gun	Immobile	40"/100cm	2	13	3+	Gun shield, Heavy anti-aircraft, Turntable.
Luftwaffe 8.8cm FlaK36 gun	Immobile	40"/100cm	2	13	3+	Gun shield, Heavy anti-aircraft, Turntable.
Firing bombardments		88"/220cm	-	3	5+	
8.8cm PaK43/41 gun	Immobile	40"/100cm	2	16	3+	Gun shield.
8.8cm PaK43 gun	Immobile	40"/100cm	2	16	3+	Gun shield, Turntable.
7.5cm GebG36 gun	Heavy	16"/40cm	2	9	3+	Gun shield, Smoke.
Firing bombardments		72"/180cm	-	3	6	Smoke bombardment.
7.5cm FK18 gun	Heavy	24"/60cm	2	9	3+	Gun shield, Smoke.
Firing bombardments		64"/160cm	-	3	6	Smoke bombardment.
7.62cm FK288(r) gun (ZIS-3)	Heavy	32"/80cm	2	9	3+	Gun shield.
Firing bombardments		80"/200cm	-	3	6	
7.62cm IKH290(r) gun (obr 1927)	Light	16"/40cm	2	5	3+	Gun shield.
Firing bombardments		64"/160cm	-	3	6	
10.5cm LG40 recoilless gun	Light	16"/40cm	1	10	2+	Breakthrough gun, Gun shield, Recoilless, Smoke.
Firing bombardments		64"/160cm	-	4	4+	
10cm leFH14/19(t) howitzer (100/17)	Immobile	24"/60cm	1	10	2+	Breakthrough gun, Gun shield, Smoke.
Firing bombardments		72"/180cm	-	4	4+	Smoke bombardment.
10.5cm leFH325(f) howitzer	Heavy	16"/40cm	1	7	2+	Breakthrough gun, Gun shield, Smoke.
Firing bombardments		72"/180cm	-	4	4+	Smoke bombardment.

Weapon	Mobility	Range	ROF	Anti-tank	Firepower	Notes
10.5cm GebH40 howitzer	Heavy	24"/60cm	1	10	2+	Breakthrough gun, Smoke.
Firing bombardments		72"/180cm	-	4	4+	Smoke bombardment.
10.5cm leFH18/40 howitzer	Heavy	24"/60cm	1	10	2+	Breakthrough gun, Gun shield, Smoke.
Firing bombardments		72"/180cm	-	4	4+	Smoke bombardment.
10.5cm leFH18 howitzer	Immobile	24"/60cm	1	10	2+	Breakthrough gun, Gun shield, Smoke.
Firing bombardments		72"/180cm	-	4	4+	Smoke bombardment.
s10cm K18 gun	Immobile	32"/80cm	1	15	2+	
Firing bombardments		96"/240cm	-	4	4+	
12.2cm FH396(r) howitzer	Immobile	24"/60cm	1	8	2+	Breakthrough gun, Gun shield.
Firing bombardments		80"/200cm	-	4	3+	
12.8cm K81 gun	Immobile	48"/120cm	1	17	2+	Breakthrough gun.
Firing bombardments		104"/260cm	-	4	3+	
15cm sFH18 howitzer	Immobile	24"/60cm	1	13	1+	Bunker buster, Smoke.
Firing bombardments		80"/200cm	-	5	2+	Smoke bombardment.
15.5cm sFH414(f) howitzer	Immobile	16"/40cm	1	10	1+	Bunker buster, Gun shield.
Firing bombardments		72"/180cm	-	5	2+	
15cm NW41 rocket launcher	Light	64"/160cm	-	3	4+	Rocket launcher, Smoke bombardment.
21cm NW42 rocket launcher	Light	72"/180cm	-	3	3+	Rocket launcher, Smoke bombardment.

INFANTRY TEAMS

Team	Range	ROF	Anti-tank	Firepower	Notes
Rifle team	16"/40cm	1	2	6	
Rifle/MG team	16"/40cm	2	2	6	
MG team	16"/40cm	3	2	6	ROF 2 when pinned down
SMG team	4"/10cm	3	1	6	Full ROF when moving.
Assault Rifle team	8"/20cm	3	1	6	Full ROF when moving.
Light Mortar team	16"/40cm	1	1	4+	Can fire over friendly troops.
Tank-hunter team	8"/20cm	1	11	5+	Tank Assault 5.
Panzerschreck team	8"/20cm	2	11	5+	Tank Assault 5.
Flame-thrower team	4"/10cm	2	-	6	Flame-thrower.
Staff team	16"/40cm	1	2	6	Moves as a Heavy Gun team.
Feldgendarm team	4"/10cm	3	1	6	Full ROF when moving.

Additional Training and Equipment

Panzerfaust	4"/10cm	1	12	5+	Tank Assault 6, Cannot shoot in the Shooting Step if moved in the Movement Step.

Panzerknacker teams are rated as Tank Assault 5.

Pioneer teams are rated as Tank Assault 4.

TRANSPORT

Vehicle *Weapon*	Mobility *Range*	Armour Front *ROF*	Armour Side *Anti-tank*	Armour Top *Firepower*	Equipment and Notes
Trucks					
BMW motorcycle & sidecar or Kübelwagen jeep	Jeep	-	-	-	Optional Passenger-fired hull MG
Schwimmwagen	Jeep	-	-	-	Amphibious, Passenger-fired hull MG.
Horch, Krupp, or Steyr Kfz 70, Opel Blitz 3-ton, or Opel Kfz 68 radio truck	Wheeled	-	-	-	
Kettenkrad half-track or Horch Kfz 15 car	Jeep	-	-	-	
Opel Maultier	Half-tracked	-	-	-	
Horse-drawn wagon	Wagon	-	-	-	
Reinforced Truck	Wheeled	-	-	-	Passenger-fired hull MG, Overloaded, Unreliable.

Tractors

Sd Kfz 10, Sd Kfz 11, or Sd Kfz 7 half-track	Half-tracked	-	-	-	
Horse-drawn limber	Wagon	-	-	-	
RSO	Slow Tank	-	-	-	

Recovery Vehicles

Sd Kfz 9 Famo half-track	Half-tracked	-	-	-	Recovery vehicle.
Bergepanzer III recovery vehicle	Standard Tank	5	3	0	AA MG, Recovery vehicle.
Bergepanther recovery vehicle	Standard Tank	10	5	0	AA MG, Wide tracks, Recovery vehicle.
Bergepanzer Elefant Recovery Vehicle	Slow Tank	9	8	2	Hull MG, AA MG, Unreliable, Recovery vehicle

Armoured Personnel Carriers

Sd Kfz 250/1 or 251/1 half-track	Half-tracked	1	0	0	Hull MG, Passenger-fired AA MG.
Sd Kfz 250/1 (HMG) or 251/1 (HMG) half-track	Half-tracked	1	0	0	Hull MG, HMG Carrier, Passenger-fired AA MG.
Sd Kfz 250/10 (3.7cm) half-track	Half-tracked	1	0	0	Passenger-fired AA MG.
3.7cm PaK36	*16"/40cm*	*2*	*6*	*4+*	*Hull mounted.*
Sd Kfz 250/11 (2.8cm) half-track	Half-tracked	1	0	0	Passenger-fired AA MG.
2.8cm sPzB41	*16"/40cm*	*2*	*7*	*5+*	*Hull mounted, No HE.*
Sd Kfz 250 (2cm) half-track	Half-tracked	1	0	0	Passenger-fired AA MG.
2cm Flak38	*16"/40cm*	*3*	*5*	*5+*	*Hull mounted.*
Sd Kfz 251/1 (2.8cm) half-track	Half-tracked	1	0	0	Passenger-fired AA MG.
2.8cm sPzB41	*16"/40cm*	*2*	*7*	*5+*	*Hull mounted, No HE.*
Sd Kfz 251/10 (3.7cm) half-track	Half-tracked	1	0	0	Passenger-fired AA MG.
3.7cm PaK36	*16"/40cm*	*2*	*6*	*4+*	*Hull mounted.*
Sd Kfz 251/7 (Pioneer) half-track	Half-tracked	1	0	0	Hull MG, Passenger-fired AA MG, Assault bridge.
Sd Kfz 251/17 (2cm) half-track	Half-tracked	1	0	0	Passenger-fired AA MG.
2cm FlaK38	*16"/40cm*	*3*	*5*	*5+*	*Self-defence anti-aircraft. (Earth & Steel)*
Sd Kfz 251/1 (Stuka) half-track	Half-tracked	1	0	0	Hull MG, Passenger-fired AA MG.
28cm sW40 Rocket Launcher	*40"/100cm*	*-*	*3*	*1+*	*Hull mounted, Stuka zu Fuss.*
U304(f) leSPW half-track	Half-tracked	1	0	0	Hull MG, Passenger-fired AA MG, Unreliable.
U304(f) (HMG) half-track	Half-tracked	1	0	0	Hull MG, HMG carrier, Passenger-fired AA MG, Unreliable.
U304(f) (3.7cm)	Half-tracked	1	0	0	Passenger-fired AA MG, Unreliable.
3.7cm PaK36 gun	*16"/40cm*	*2*	*6*	*4+*	*Hull mounted.*
S307(f) (Pioneer) half-track	Half-tracked	1	0	0	Hull MG, Passenger-fired AA MG, Assault bridge, Unreliable.

AIRCRAFT

Aircraft	Weapon	To Hit	Anti-tank	Firepower	Notes
Ju 87D Stuka	Bombs	4+	5	1+	
Ju 87G Stuka	Cannon	3+	11	4+	
Hs 129B	Cannon	2+	9	4+	Flying tank.
Hs 129B3	Cannon	4+	15	3+	Flying tank, No HE.
	MG	3+	6	5+	
Bf 109E	Cannon	3+	7	5+	
	Bombs	4+	5	2+	
Fw 190F	Cannon	3+	7	5+	
	Bombs	4+	5	1+	

BRITAIN

NATIONAL RULES

DOCTRINE

British Bulldog

When British platoons test their Motivation to Counterattack in assaults, you may re-roll the die and apply the re-rolled result to all British platoons that failed on the original roll.

Carry On, Sergeant

A British platoon may Charge into Combat and Consolidate without having a Platoon Command team, and if it does so, the Assaulting teams do not need to be or remain In Command.

Night Attack

A player commanding a British Infantry Company may elect to make a Night Attack in missions that do not use the Meeting Engagement special rule (see page 264 of the rulebook). If they do so, the game starts at night and uses the Dawn rules on page 273 of the rulebook.

When making a Night Attack, the company has the Always Attacks special rule (see page 257 of the rulebook), and all British platoons containing only Infantry teams can use the Spearhead Deployment rule (see page 261 of the rulebook).

ANTI-TANK GUNS

Tip and Run

A British Platoon with the Tip and Run special rule may attempt to Tip and Run immediately after shooting.

Roll a Skill Test for the platoon.

- If the test is passed, the platoon treats the remainder of the Shooting Step as a Movement Step in which it can move up to another 4"/10cm away from all enemy platoons that it shot at.
- If the test is failed, the platoon remains where it is.

Either way, a platoon that attempts to Tip and Run cannot take any part in an assault in the same turn.

Tow Hooks

Any tank fitted with a tow hook may tow a 6 pdr or 17 pdr gun, carrying the crew as Passengers Riding on Tanks (see page 47 of the rulebook).

TANKS

Semi-indirect Fire

Weapons capable of Semi-indirect Fire that didn't move in the Movement Step may re-roll failed rolls to hit when shooting their main guns at platoons with all teams more than 16"/40cm away.

ARTILLERY

Eight-gun or Twelve-Gun Batteries

British Artillery Batteries with Staff teams use the Independent Command, Combined Bombardment, and Mike Target special rules.

Combined Bombardment

The Gun Troop with the battery's Staff team, or any other Gun Troop within Command Distance of the battery's Staff team, may act as the base of a Combined Bombardment using the Mixed Bombardment rule on page 131 of the rulebook. Any other Gun Troops from the battery (whether or not they are near the Staff team) may join the base troop before rolling to Range In and fire as a single combined Artillery Battery.

If a Gun Troop from another battery that has the Combined Bombardment special rule has a Staff team or is within Command Distance of its Staff team, it may also be combined into the Bombardment, along with any other Gun Troops from its battery.

HMG Bombardments

British Vickers HMG teams, and HMG Carriers firing as HMG teams, can fire bombardments (unless their platoon entry specifically says otherwise). These are conducted in the same way as any other Artillery Bombardment. However, since they have Anti-tank and Firepower ratings of '-', they cannot harm Armoured vehicles or Destroy teams in Bulletproof Cover, although they can still Pin them Down.

Independent Command

The battery's Staff team must be attached to a Gun Troop from its battery at the start of the game before deployment. However, any Gun Troop within Command Distance of the battery's Staff team, counts as having a Staff team (allowing it to use the All Guns Repeat! rule).

If the battery has an HQ Troop Command team, it must also be attached to a Gun Troop from its battery at the start of the game before deployment, becoming the Platoon Command team for the Gun Troop. If the HQ Troop Command team is Destroyed, the original Platoon Command team resumes that role.

Mike Target

A battery may re-roll their first failed attempt to Range In. If they fail on the re-roll, they still have two more ranging attempts left as normal. They do not get a re-roll on their second and third attempts.

When a battery repeats a Bombardment using the All Guns Repeat! Rule (see page 128 of the rulebook), the entire regiment joins in. As a result any teams caught under the Bombardment must re-roll successful Saves.

Some rules refer to Bombardments using the Mike Target special rule as Stonks or Murders.

CANADIAN

Assault Troops

Canadian Platoons do not use the British Bulldog special rule. Instead any Canadian Platoon may re-roll failed Motivation tests to Rally from being Pinned down or to remount vehicles after being Bailed Out.

Woodsmen

Canadian Platoons use the German Mission Tactics special rule (see page 242 of the rulebook).

GUARDS

Unflappable

Guards platoons fight to the last and may re-roll any failed Platoon Morale Checks and Company Morale Checks.

TURNING TIDE

6TH AIRBORNE DIVISION

Section Mortars

Each turn one of the Parachute Platoon's Rifle/MG teams may fire as a Light Mortar team firing smoke.

Gammon Bombs

All teams carrying Gammon Bombs have Tank Assault 3.

RIFLE COMPANY (50TH, 3RD AND CANADIAN 3RD DIVISIONS)

Dead Men Walking

The Company Command Team from the British 50th and 3rd Division and the Canadian 3rd Division may re-roll the first Company Morale Check that they are required to take if they fail their first attempt.

King's Own Scottish Borderers

You may add a bagpiper to the 2iC Command team of a Rifle Company from the British 3rd Division for +10 points.

If the 2iC Command team is Destroyed by enemy shooting, the enemy rolls a die to destroy the 2iC Command team using the Warrior Infantry Team Casualties rule on page 106 of the rulebook. Instead of requiring a 4+ to destroy the 2iC Command team, the enemy player needs a roll of 5+ to destroy the team. If the enemy's roll is unsuccessful, the piper's music inspires the surviving team members to keep going and the team is returned to play immediately.

Any hits on the 2iC Command team do not count towards Pinning Down the platoon.

79TH ARMOURED DIVISION, BREACHING GROUPS

A Breaching Group is unusual in that it's composition can change from game to game. A Breaching Group may contain multiple platoons and they do not all have to be equipped the same. These include: AVRE Sections; a Field Platoon, Royal Engineers; a Sherman Crab Platoon; and Armoured Personnel Carrier Sections.

The Breaching Group is made up of Engineering teams. There is no actual team called an Engineering team. Instead before each game these are exchanged for engineering platoons listed below.

Field Platoon, Royal Engineers

Exchange an Engineering team for six Pioneer Rifle teams. You may exchange up to two Engineering teams for Pioneer Rifle teams. Add an additional Command Pioneer Rifle team and group all Pioneer Rifle teams together to form a single Field Platoon, Royal Engineers. Your force may contain up to one Field Platoon, Royal Engineers. At the beginning of the game before deployment you may replace three Pioneer Rifle teams with a Pioneer Supply Truck.

Specialist Troops

Field Platoons, Royal Engineers may not launch assaults, nor may they counterattack if assaulted. Field Platoons, Royal Engineers will always attempt to Break Off at the earliest opportunity.

Sherman Crab Platoon

Exchange one Engineering team for a Sherman Crab flail tank. You may exchange up to four Engineering teams for Sherman Crab flail tanks. Group all Sherman Crab flail tanks together to form a single Sherman Crab Platoon. Nominate one Sherman Crab flail tank to become the Platoon Command tank. Your force may contain up to one Sherman Crab Platoon.

Engineering Equipment

Sherman Crab platoons may not launch assaults.

Mine Flail

Sherman Crab Platoons use the rules for Mine Flails found on page 226 of the rulebook.

AVRE Sections

Exchange one Engineering team for two AVRE assault tanks to form an AVRE Section. Nominate one AVRE assault tank to become the section Command team. Your force may contain up to three AVRE Sections.

Pioneers

At the start of the turn you may elect to operate an AVRE crew dismounted. If you do this, the AVRE cannot move, shoot, or assault, but can undertake engineering tasks (see pages 225 to 227 of the rulebook) as if it was an infantry Pioneer team. Until the start of your next turn, the enemy may elect to count the AVRE as an Infantry team for any Saves it must take from shooting or assaults. Place a spare pioneer or rifle team beside the AVRE as a marker to indicate that it is operating dismounted.

Fascine and SBG Bridge

At the start of the game after fortifications (if any) are placed, but before deployment, you may elect to mount a fascine or SBG bridge on any or all of your AVsRE. An AVRE that has a fascine or SBG bridge mounted cannot shoot. In addition an AVRE that has an SBG bridge mounted is Overloaded rather than Wide-tracked. Fascines and SBG bridges are Assault bridges (see page 226 of the rulebook). Fascines may only be used to bridge ditches and craters. SBG bridges can be used to bridge these as well as streams and other gaps and to create a ramp for crossing an anti-tank wall, a sea wall, or a bank up to 1 1/2"/4cm high. An AVRE may abandon its fascine or SBG bridge instead of shooting. Remove the fascine or bridge from play.

Engineering Equipment

AVRE Sections may not launch assaults.

Top Armour

Any Armoured vehicle hit by a Petard mortar uses its Top armour rating for its Armour Saves.

Anti-tank Obstacles

Instead of shooting, an AVRE may gap anti-tank walls and dragon's teeth. These are both forms of Anti-tank Obstacle (see page 231 of the rulebook). The AVRE must start the turn within 4"/10cm and Line of Sight of the obstacle. In the Shooting Step the AVRE fires at the obstacle. Roll a Skill Test. If the test is successful, the Anti-tank Obstacle is removed and the area becomes Difficult Going, otherwise, the obstacle remains intact.

Armoured Personnel Carrier Sections

Exchange one Engineering team to add two Defrocked Priest Armoured Personnel Carriers (APC) to each Rifle Platoon in your company (including Rifle Platoons taken as support). These APCs are part of the Rifle Platoon for all purposes. Your Company HQ travels in any of the platoons' vehicles.

Deploying APCs

If you equip your Rifle Platoons in you Rifle company with Armoured Personnel Carriers, your force becomes a Mechanised Company. You may still conduct Night Attacks with APCs.

Engineering Equipment

Defrocked Priests may not launch assaults.

APCs and Passengers

Defrocked Priest APCs are Armoured Transport teams.

Morale Checks

As Transport teams Defrocked Priest APCs are ignored in Platoon Morale Checks.

Job Done!

At the beginning of the game before deployment, you may exchange all four Engineering teams of the Breaching Group as follows:

DD Armoured Platoon: Exchange all four Engineering teams for an Independent Armoured Platoon with three Sherman V tanks and one Firefly VC tank (see page 121 of *Turning Tide*).

Corps Anti-tank Platoon (SP), Royal Artillery: Exchange all four Engineering teams for a Corps Anti-tank Platoon (SP), Royal Artillery with four M10C 17 pdr (SP) guns (see page 120 of *Turning Tide*).

Your force may not contain more than one platoon equipped with M10C 17 pdr or M10 vehicles.

COMMANDO – BEACHES AND ORNE

No British Bulldog

The British Bulldog rule does not apply to Commando platoons

Fairburn-Sykes

Commando Infantry Teams hit on a roll of 2+ in an assault.

Assault Force

A force containing a Commando (Beaches) Always Attacks against any other Infantry Company. A force containing a Commando (Orne) is not affected by the Assault Force special rule, and determines who attacks as normal in missions.

Know the Plan

Commando Platoons use the German Mission Tactics special rule (see page 242 of the rulebook).

Mind and Heart

All commando Infantry and Man-packed Gun teams are Mountaineers (see page 61 of the rulebook).

You Are Not Alone

A Commando never has to take a Company Morale Check for having more platoons destroyed than still fighting. They take Platoon Morale Checks as normal though. However, if the total number of Commando Sections Destroyed from the Commando exceeds the number still on the table, the Commando player must roll a die at the start of each turn. On a roll of 3+, the Commando fights on heedless of its losses. On any other roll, the game ends. If the Commando no longer has a 2iC or Company Command team, then this roll is automatically failed.

NORMANDY GUARDS TANK SQUADRON

You may field a Normandy Guards Tank Squadron representing a force from the 6th Guards Tank Brigade. If you field this force, your Company HQ and all your Combat Platoons are Guards Platoons at an additional cost of +10 points per platoon.

Normandy Guards Platoons use the Unflappable special rule in the British National Rules section above.

7TH ARMOURED DIVISION

Cautious Not Stupid

A platoon from the 7th Armoured Division that did not fail any Saves in the previous enemy Shooting Step, may re-roll any failed attempts to Rally Pinned Down platoons or Remount Bailed Out vehicles.

10CIB & SOUTH ALBERTA REGIMENT

A Little Help Here

If the Command teams of a Rifle Platoon and an Armoured Recce Platoon are within Command Distance of each other at the start of the Shooting Step, any tank in the Armoured Recce Platoon that is adjacent to an Infantry team from the Rifle Platoon may re-roll misses with its machine-guns if it shoots at the same enemy platoon as the Infantry team.

Tip of the Spear

An Armoured recce Squadron or Lorried Rifle Company from the Canadian 4th Armoured Division Always Attacks.

Reconnaissance Men

Teams from the Canadian Armoured Recce Squadron HQ and Canadian Armoured Recce Platoons use the Cautious Movement rule found on page 193 of the rulebook as though they were Recce teams.

LT COLONEL R.G. PINE-COFFIN

Pine-Coffin is a Warrior Higher Command SMG team rated as Fearless Veteran.

Pine-Coffin can join a Parachute Company for +50 points.

Counterattack Platoon

At the end of any Starting Step in which the 2iC Command team is within 6"/15cm of Pine-Coffin you may form, or supplement, the Counterattack Platoon. To do so nominate up to two teams from any Parachute Platoon within 6"/15cm of the 2iC team. You may never nominate teams from a platoon of less than five teams. These teams become part of the Counterattack Platoon. The 2iC Command team becomes the Platoon Command team for the Counterattack Platoon. Teams placed in the Counterattack Platoon are no longer part of their original platoons. The Counterattack Platoon counts as a platoon for all purposes except Company Morale Checks.

CAPTAIN JOHN TRELEAVEN

Treleaven is a Canadian, a Warrior and a Platoon Command Rifle/MG team. He is rated as Fearless Veteran.

Treleaven may join a Rifle Company from 3rd Canadian Division replacing the normal Platoon Command team of a Rifle Platoon for +25 points.

Croix de Guerre

Captain Treleaven may take Company Morale Checks as if he were the Company Command team if both the Company Command team and 2iC Command team are Destroyed.

In the First Wave

Captain Treleaven and any platoon that he is leading always pass Motivation Tests (including any Company Morale Check he is required to take) on a roll of 3+.

Captain John Smith

You may field Captain Treleaven as a British Warrior with a different name. If so, he joins a Rifle Company from the British 3rd Division replacing a Rifle Platoon Command team, but is otherwise the same as Captain Treleaven.

CSM STAN HOLLIS, VC

CSM Stan Hollis is a Warrior Command SMG team. He is rated Fearless Veteran. CSM Stan Hollis can take command of a Rifle Platoon from the 50th Division during the game.

CSM Stan Hollis can join a 50th Division Rifle Company for +35 points.

HANDY WITH A GRENADE

Hollis hits on a roll of 2+ in assault combat.

FIGHTING SPIRIT

Hollis and any Rifle Platoon he is currently commanding pass Motivation tests on a roll of 3+.

BRIGADIER THE LORD LOVAT

The Lord Lovat is a Warrior Higher Command Rifle team. He is rated as Fearless Veteran. Lord Lovat is armed with his old Winchester hunting rifle.

Lovat may join a Commando that does not contain Peter Young for +25 points.

BILL MILLIN

Any hits on Lovat do not count towards Pinning Down the platoon he has joined.

IN THE NICK OF TIME

In missions that use the Reserves or Delayed Reserves special rules, the Commando player may roll one more die in addition to the normal allotment to see if reserves arrive. This additional die may only be used for Commando Platoons held in reserve.

LT COLONEL PETER YOUNG

Peter Young is a Warrior Higher Command Rifle team. He is rated Fearless Veteran. Young is armed with a M1 Garand rifle. He has a Range of 16"/40cm, ROF 1 with no penalty for moving, an Anti-tank rating of 2, and a Firepower rating of 6.

Peter Young may join a Commando that does not contain The Lord Lovat for +50 points.

MC and Double Bar

Young and any Commando, Commando Machine-gun or Commando Mortar platoon he is currently leading pass all Motivation Tests on a roll of 2+ instead of their normal roll.

Contrary

Young and any Commando, Commando Machine-gun or Commando Mortar platoon he is currently leading may re-roll any failed Skill Test to cross Impassable Terrain using the Mind and Heart rule.

Cornfields Stop Bullets

Young and any Commando, Commando Machine-gun or Commando Mortar platoon he joins can be Gone to Ground when shooting, as long as they are Concealed by terrain and did not move.

MAJOR DAVID V CURRIE, VC

Major Currie is a Warrior Company Command team mounted in a Sherman V tank and is rated as Fearless Veteran.

Major Currie replaces a Canadian 4th Armoured Division Armoured Recce Squadron Company Command Sherman V for +55 points or Rifle Company Command Rifle team for +105 points.

Calm Under Pressure

Currie may re-roll failed Company Morale Checks.

There for His Men

Once per turn, a Canadian platoon that Currie has not joined that is within Command Distance and Line Of Sight of Currie may re-roll one failed Motivation Test. This is in addition to any Company Command team Motivation test re-rolls he may make for platoons he has joined.

SERGEANT TOM STANLEY

Sergeant Tom Stanley is a Warrior OQF 6 pdr gun team and is accompanied by a Loyd Carrier. He is rated Fearless Veteran.

He replaces one OQF 6 pdr gun and Loyd Carrier in a Motor Anti-tank Platoon or Lorried Anti-tank Platoon from the 7th Armoured Division for +25 points. Stanley is no longer part of that platoon.

TOM, DICK, 'N 'ARRY

Sergeant Tom Stanley may only Join the Anti-tank Platoon from which he took his gun and carrier. He may not Join any other platoon.

Although Stanley is an Warrior team he can deploy in Ambush with his Anti-tank Platoon if it deploys in Ambush. He does not need to be placed at the same time, nor in the same location, as the rest of the platoon.

MACHINIST

Stanley re-rolls all missed rolls to hit when shooting.

MILITARY MEDAL

Stanley is a Warrior Gun team.

HELL'S HIGHWAY

BRITISH

Unflappable

Guards Platoons may re-roll any failed Platoon Morale Checks. The Company Command team of a Guards Squadron or Company may re-roll Company Morale Checks.

Platoon, Debus!

Infantry teams from Guards Rifle Platoons riding on Guards Armoured or Guards Armoured Recce Platoons as Tank-riders may Disengage from Shooting as if they were Recce Teams (see page 194 of the rulebook).

Infantry teams from a Guards Rifle Platoon do not need to Reorganise after disengaging.

The 2nd Household Cavalry

Unflappable

Platoons from a 2nd Household Cavalry Squadron may re-roll any failed Platoon Morale Checks. The Company Command team of a 2nd Household Cavalry Squadron may re-roll Company Morale Checks.

Sangfroid

An Armoured Car Platoon of the 2nd Household Cavalry may attempt to Disengage even if it shot in its previous turn (see page 194 of the rulebook).

Lieutenant Colonel J.O.E. Vandeleur, DSO & Bar

Lieutenant Colonel JOE Vandeleur is a Warrior mounted in a Humber scout car, and is a Higher Command team rated as Confident Veteran. Vandeleur may join an Irish Guards Armoured Squadron or an Irish Guards Rifle Company for +50 points. Add an RAF Forward Air Controller team for +10 points.

Guns, Left!

A force including Vandeleur does not use the Semi-indirect Fire special rule. Instead, Sherman V and Firefly VC tanks that moved in their Movement Step may re-roll failed rolls To Hit when shooting their main guns at teams up to 16"/40cm away. Guns Left! does not apply when conducting Defensive Fire.

Press On Regardless!

If Vandeleur has not moved At the Double and is not Bogged Down or Bailed Out, he can attempt to push his platoons forward using Press On Regardless.

Guards Combat, Weapons and Brigade Support Platoons that are within 8"/20cm of Vandeleur and are not Pinned Down may attempt to move again at the end of the Movement Step after all other movement. Roll a die for each platoon attempting to move again:

- If the result is 5+, the platoon may move up to another 4"/10cm, regardless of its normal movement distance.
- Otherwise, the platoon remains where it is.

The platoons may still shoot and assault as normal after making this extra move.

Platoons cannot make this extra move if they moved At the Double.

Bogged Down or Bailed Out vehicles cannot make this extra move.

Unearthly Crescendo

A Forward Air Controller team is a White scout car and an Independent Tank Team rated as Confident Veteran.

While the Forward Air Controller team is within 8"/20cm of Vandeleur or a Company Command team, all air attacks against targets within Line of Sight of the Forward Air Controller are under its control.

Aircraft under the control of a Forward Air Controller team may re-roll a failed attempt when rolling to Range In an air attack.

Unlike normal air attacks that cannot be within 16"/40cm from any friendly teams (see Safety Distance rule on page 184 of the rulebook), aircraft under the control of a Forward Air Controller will only abort if friendly teams are within 12"/30cm of the Aircraft model.

1st AIRBORNE DIVISION

SECTION MORTARS

Each turn one of the Parachute Platoon's Rifle/MG teams may fire as a Light Mortar team firing smoke.

GLIDER PILOT PLATOON MISSION TACTICS

A Glider Pilot Platoon uses the German Mission Tactics special rule (see page 242 of the rulebook).

BODY ARMOUR

All SMG, Mortar and PIAT teams in the Glider Pilot Platoon are equipped with body armour. When a team with body armour is hit by an artillery bombardment and fails its save, roll a die after all Firepower Tests have been made (if any apply).

- On a result of 6, the team's body armour has saved it from harm and is not Destroyed.
- Otherwise, the team is Destroyed as normal.

AIRBORNE RECONNAISSANCE SQUADRON

Teams of the Airborne Reconnaissance Squadron HQ and Airborne Reconnaissance Platoons are not Reconnaissance teams. However, they may make a Reconnaissance Deployment move (see page 193 of the rulebook) as though they were Reconnaissance teams.

MEDIUM ARTILLERY SUPPORT

If an Airlanding Light Battery provided with Medium Artillery Support fires an All Guns Repeat! Bombardment, it fires as though it has an Anti-tank rating of 5 and a Firepower of 2+.

THE 2nd HOUSEHOLD CAVALRY

UNFLAPPABLE

Platoons from a 2nd Household Cavalry Squadron may re-roll any failed Platoon Morale Checks. The Company Command team of a 2nd Household Cavalry Squadron may re-roll Company Morale Checks.

SANGFROID

An Armoured Car Platoon of the 2nd Household Cavalry may attempt to Disengage even if it shot in its previous turn (see page 194 of the rulebook).

FROST'S PERIMETER FORCE

Frost's Perimeter Force starts the game in Prepared Positions. In addition, it is always the defender against another Infantry Company in missions that use the Defensive Battle or Mobile Battle special rules.

LIEUTENANT COLONEL JOHN FROST

Lieutenant Colonel John Frost is a Warrior Higher Command SMG team rated as Fearless Veteran.

Lieutenant Colonel John Frost may join a Parachute Company for +65 points.

FROST'S PERIMETER!

Before deployment choose one platoon in your company that is not placed in Reserve. Do not deploy this platoon when you would normally. Instead, it is deployed at the same time as Independent Teams.

MAN THE GUNS!

When one of your gun teams is Destroyed as a result of enemy shooting, roll a die. On a 5+ you may instead Destroy another friendly Infantry team within 6"/15cm of that gun team.

PERIMETER TROOPS

A force led by Frost may form a Perimeter Platoon. At the start of the game before any platoons are Deployed and before any Combat Attachments are made, a Force led by Frost with a 2iC Command team may take up to half of the teams (counting the Platoon Command team) from any Combat or Weapons platoons in the company and place them in a special Perimeter Platoon. The 2iC Command team is the Command team for this platoon.

You must leave at least two Tank teams or three other teams in each of the Combat or Weapons platoons you take teams from. Teams placed in the Perimeter Platoon are no longer part of their original platoons.

If a platoon attaching troops to the Perimeter Platoon has Transport teams, you must take the Transport team that normally carries a team into the Perimeter Platoon with it. If that Transport team normally carries other teams as well, they must join the Perimeter Platoon as well.

The Perimeter Platoon may include any HQ Support Weapons (including Infantry teams). Any remaining HQ Support Weapons must be attached out as normal.

The Perimeter Platoon has the lowest Motivation rating and the lowest Skill rating of any of the teams in it.

A HEROIC DEFENCE

Frost, and any platoon he is currently leading, always pass all Motivation Tests on a roll of 2+.

Dixie Air Support

At the start of your turn, you may elect to either use your Auster AOP as normal (using the Air Observation Post rules on page 139 of the rulebook) or use it to coordinate Dixie Air Support. The Auster AOP cannot do both in the same turn.

If you choose to use the AOP to coordinate Dixie Air Support, all air attacks against targets within Line of sight and 16"/40cm of the Auster AOP are under its control.

Aircraft under the control of an Auster AOP may re-roll a failed attempt to Range In.

Unlike normal air attacks that cannot be within 16"/40cm of any friendly teams (see Safety Distance rule on page 184 of the rulebook), aircraft under the control of an Auster AOP will only abort if friendly teams are within 12"/30cm of the Aircraft model.

4TH & 8TH INDIAN DIVISIONS

War Cry

Indian and Gurka platoons do not use the British Bulldog special rule. Instead, when enemy platoons first test their Motivation to Counterattack Indian or Gurka platoons in assaults, they must re-roll the die and apply the re-rolled result to all the platoons that passed on the original roll.

North-west Frontier

All Indian Infantry and Man-packed Gun teams are Mountaineers.

GURKHA

Khukuri

If there are no enemy Tank teams or Bunkers within 2"/5cm of a Gurka team, the Gurka team hits on a roll of 2+ in Assaults.

2ND NEW ZEALAND DIVISION

4 by 2 and No. 8 Wire

New Zealand Platoons use the German Mission Tactics special rule (see page 242 of the rulebook).

Rumpus

If a Field Battery, Royal Artillery (but not a Medium Battery, Royal Artillery) taken in support of a New Zealand Rifle Company fires a Mike Target bombardment (see page 248 of the rulebook), and ranges in on their first attempt (with or without the re-roll), each team hit by the bombardment must re-roll successful saves in the first turn of the bombardment. If a Medium Battery, Royal Artillery joins the Bombardment, it is no longer a Rumpus Bombardment as they are not New Zealand artillery.

Thousand Man Limit

New Zealand companies pass their Company Morale Checks on a role of 5+.

MAORI

Haka

When enemy platoons first test their Motivation to Counterattack Maori platoons in assaults, they must re-roll the die and apply the re-rolled result to all the platoons that passed on the original roll.

WEB INTELLIGENCE BRIEFINGS

SCOTS DIVISIONS

BAGPIPES

If the 2iC Command team is Destroyed by enemy shooting, the enemy rolls a die to destroy the 2iC Command team using the Warrior Infantry Team Casualties rule on page 106 of the rulebook. Instead of requiring a 4+ to destroy the 2iC Command team, the enemy player needs a roll of 5+ to destroy the team. If the enemy's roll is unsuccessful, the piper's music inspires the surviving team members to keep going and the team is returned to play immediately.

Any hits on the 2iC Command team do not count towards Pinning Down the platoon.

BREACHING GROUP

BREACHING GROUP

You can find the Breaching Group rules in the Turning Tide section above.

JOB DONE!

At the beginning of the game before deployment, you may exchange all four Engineering teams of the Breaching Group as follows:

Armoured Platoon: Exchange all four Engineering teams for an Armoured Platoon with three Sherman I or II tanks and one Firefly VC tank. This Armoured Platoon is rated Confident Trained.

51ST HIGHLAND DIVISION

CAUTIOUS NOT STUPID

A platoon from the 51st Highland Division that did not fail any Saves in the previous enemy Shooting Step, may re-roll any failed attempts to Rally Pinned Down platoons or Remount Bailed Out vehicles.

GRENADIER GUARDS

UNFLAPPABLE

Guards platoons fight to the last and may re-roll any failed Platoon Morale Checks and Company Morale Checks.

11TH ARMOURED DIVISION

DEVIL'S OWN

Armoured Car and Scout Car Patrols may re-roll failed Skill Tests to Disengage (see page 194 of the rulebook).

CAPTAIN GEOFFREY WALTER FOWNES LUTTRELL, MC

Captain Luttrell is a Warrior mounted in a Command Cromwell VI CS tank and rated as Confident Trained. Luttrell is the Command team of Luttrell's Close Support Platoon.

Luttrell may join an Armoured Recce Squadron of the 15th/19th Hussars. An Armoured Recce Squadron including Luttrel may not purchase Cromwell VI CS tanks in the Armoured Recce Squadron HQ.

Luttrell's Command Cromwell VI CS with:

1 Cromwell VI CS	110 points
3 Cromwell VI CS	210 points
5 Cromwell VI CS	310 points

Regimental Gunnery Officer

Any artillery bombardment that includes Luttrell may reroll its first attempt to Range In.

Recovery Expert

Luttrell and any team in a platoon he is currently commanding may reroll failed attempts to free themselves from being Bogged Down.

MAJOR GENERAL ROY URQUHART

Major General Roy Urquhart is a Warrior Higher Command Pistol team rated Fearless Veteran. Major General Roy Urquhart may join a British Parachute or an Airlanding Company (found in *A Bridge Too Far*) for +25 points.

To The Bridge

Infantry and Man-packed Gun teams in a platoon led by Urqhart may move 16"/40cm instead of the normal 12"/30cm when moving At the Double.

Establish the Perimeter!

A platoon led by Major General Urquhart may re-roll failed attempts to Dig In.

Licensed to Kill

Urquhart may re-roll failed To Hit rolls with his pistol and does not suffer the normal +1 To Hit penalty when Moving with ROF 1.

LIEUTENANT COLONEL SANDY THOMAS

Lieutenant Colonel Sandy Thomas is a Warrior Higher Command Rifle team rated Fearless Veteran. Lieutenant Colonel Sandy Thomas may join a New Zealand Rifle Company (found in *Cassino*) for +55 points.

YOU MISSED ME!

Your opponent must re-roll successful results to Destroy Thomas using the Warrior Infantry Team Casualties rule.

SEND THEM IN!

Once each turn you may re-roll one die rolled to receive Reserves for your force.

DECEPTION

At the start of the game when you would normally deploy Warrior and Independent teams, you may re-deploy one of your Combat or Weapons platoons on the table anywhere within your deployment area.

SEPOY KAMAL RAM

Sepoy Kamal Ram is a Warrior Rifle/MG team and rated Fearless Veteran. Sepoy Kamal Ram may join an Indian Rifle Company (found in *Cassino*) for +75 points.

RECON SCOUT

Ram may use the Eyes and Ears rule (see page 195 of the rulebook) as though he were a Recce platoon.

SPEAR POINT

At the start of the game before deployment, you may place one platoon in Immediate Ambush (see page 267 of the rulebook).

KEEP YOUR HEAD DOWN

When Ram joins one of your Combat or Weapons platoons, it may use the Cautious Movement rule (see page 193 of the rulebook).

BRITISH ARSENAL

TANK TEAMS

Name *Weapon*	Mobility *Range*	Armour Front *ROF*	Armour Side *Anti-tank*	Armour Top *Firepower*	Equipment and Notes
LIGHT TANKS					
Stuart V or VI	Light Tank	4	2	1	Co-ax MG, Hull MG.
M6 37mm gun	*24"/60cm*	*2*	*7*	*4+*	
Stuart VI Jalopy	Light Tank	4	2	0	Hull .50 Cal MG, Hull MG, Wide tracks.
Tetrarch	Light Tank	1	1	1	Co-ax MG.
OQF 2 pdr gun	*24"/60cm*	*2*	*7*	*4+*	
With Littlejohn adaptor	*24"/60cm*	*2*	*9*	*5+*	*No HE.*
With OQF 3" howitzer	*24"/60cm*	*2*	*5*	*3+*	*Smoke.*
MEDIUM TANKS					
Centaur	Standard Tank	6	4	1	Co-ax MG, Protected ammo, Unreliable.
OQF 95mm CS howitzer	*24"/60cm*	*2*	*7*	*3+*	*Smoke.*
Firing bombardments	*48"/120cm*	*-*	*4*	*5+*	*Smoke bombardment.*
Cromwell IV	Light Tank	6	4	1	Co-ax MG, Hull MG, Protected ammo, Tow hook.
OQF 75mm gun	*32"/80cm*	*2*	*10*	*3+*	*Semi-indirect fire, Smoke.*
Cromwell VI CS	Light Tank	6	4	1	Co-ax MG, Hull MG, Protected ammo, Tow hook.
OQF 95mm CS howitzer	*24"/60cm*	*2*	*7*	*3+*	*Smoke.*
Firing bombardments	*48"/120cm*	*-*	*4*	*5+*	*Smoke bombardment.*
Challenger A30	Light Tank	6	4	1	Co-ax MG, Overloaded, Protected ammo, Tow hook.
OQF 17 pdr gun	*32"/80cm*	*3*	*13*	*3+*	*No HE, Semi-indirect fire.*
Sherman I, II, III, or V	Standard Tank	6	4	1	Co-ax MG, Hull MG, Tow hook.
M3 75mm gun	*32"/80cm*	*2*	*10*	*3+*	*Semi-indirect fire, Smoke.*
DD Sherman	Standard Tank	6	4	1	Co-ax MG, DD Tank.
M3 75mm gun	*32"/80cm*	*2*	*10*	*3+*	*Stabilisers, Smoke.*
Firefly VC	Standard Tank	6	4	1	Co-ax MG, Tow hook.
OQF 17 pdr gun	*32"/80cm*	*2*	*13*	*3+*	*No HE, Semi-indirect fire.*
INFANTRY TANKS					
Churchill III or IV	Slow Tank	8	7	2	Co-ax MG, Hull MG, Protected ammo, Tow hook, Wide tracks.
OQF 6 pdr	*24"/60cm*	*3*	*10*	*4+*	
Churchill IV (NA75)	Slow Tank	8	7	2	Co-ax MG, Hull MG, Protected ammo, Tow hook, Wide tracks.
M3 75mm gun	*32"/80cm*	*2*	*10*	*3+*	*Semi-indirect fire, Smoke.*
Churchill V CS	Slow Tank	8	7	2	Co-ax MG, Hull MG, Protected ammo, Tow hook, Wide tracks.
OQF 95mm CS howitzer	*24"/60cm*	*2*	*7*	*3+*	*Smoke.*
Firing bombardments	*48"/120cm*	*-*	*4*	*5+*	*Smoke bombardment.*
Churchill VI	Slow Tank	8	7	2	Co-ax MG, Hull MG, Protected ammo, Tow hook, Wide tracks.
OQF 75mm gun	*32"/80cm*	*2*	*10*	*3+*	*Semi-indirect fire, Smoke.*
Churchill VII	Slow Tank	13	8	2	Co-ax MG, Hull MG, Protected ammo, Tow hook, Wide tracks.
OQF 75mm gun	*32"/80cm*	*2*	*10*	*3+*	*Semi-indirect fire, Slow traverse, Smoke.*
Churchill Crocodile	Slow Tank	13	7	1	Co-ax MG, Protected ammo, Wide tracks.
OQF 75mm gun	*32"/80cm*	*2*	*10*	*3+*	*Slow traverse, Smoke.*
Crocodile flame-gun	*4"/10cm*	*5*	*-*	*5+*	*Hull-mounted, Flame-thrower, Fuel trailer.*
ENGINEERING TANKS					
Churchill AVRE	Slow Tank	8	7	2	Co-ax MG, Hull MG, Protected ammo, Wide-tracks.
Petard mortar	*4"/10cm*	*1*	*5*	*1+*	*Bunker buster, Demolition mortar, Slow traverse.*
Sherman Crab	Standard Tank	6	4	1	Co-ax MG, Overloaded, Mine flail.
M3 75mm gun	*32"/80cm*	*2*	*10*	*3+*	*Smoke.*

Self-propelled Anti-aircraft Guns

Crusader A/A	Standard Tank	2	2	1	Fast Tank, Unreliable.
Twin 20mm gun	*16"/40cm*	*5*	*5*	*5+*	*Anti-aircraft.*
Oerlikon 20mm SP	Wheeled	-	-	-	
Oerlikon 20mm gun	*16"/40cm*	*4*	*5*	*5+*	*Anti-aircraft.*
Polsten 20mm SP	Wheeled	-	-	-	
Polsten 20mm gun	*16"/40cm*	*4*	*5*	*5+*	*Anti-aircraft.*
Bofors 40mm SP	Wheeled	-	-	-	Gun shield, Awkward layout.
Bofors 40mm gun	*24"/60cm*	*4*	*6*	*4+*	*Anti-aircraft.*

Self-propelled Anti-tank Guns

M10 3" SP	Standard Tank	4	2	0	.50 cal AA MG.
M7 3" gun	*32"/80cm*	*2*	*12*	*3+*	*Slow traverse.*
M10C 17 pdr SP	Standard Tank	4	2	0	.50 cal AA MG.
OQF 17 pdr gun	*32"/80cm*	*2*	*13*	*3+*	*No HE, Slow traverse.*

Self-propelled Guns

Sexton	Standard Tank	1	0	0	AA MG.
OQF 25 pdr gun	*24"/60cm*	*2*	*9*	*3+*	*Hull mounted, Smoke.*
Firing bombardments	*80"/200cm*	*-*	*4*	*5+*	*Smoke bombardment.*
Priest	Standard Tank	1	0	0	.50 cal AA MG.
M2A1 105mm howitzer	*24"/60cm*	*1*	*9*	*2+*	*Hull mounted, Breakthrough gun, Smoke.*
Firing bombardments	*72"/180cm*	*-*	*4*	*4+*	*Smoke bombardment.*
Sherman OP	Standard Tank	6	4	1	Hull MG.
AEC III	Slow Wheeled	4	2	0	Co-ax MG, Overloaded.
M3 75mm gun	*32"/80cm*	*2*	*10*	*3+*	*Smoke.*
Autocar 75mm SP	Half-tracked	1	0	0	
M1897 75mm gun	*24"/60cm*	*2*	*9*	*3+*	*Hull-mounted, Smoke.*
Firing bombardments	*64"/160cm*	*-*	*2*	*6*	*Smoke bombardment.*

Infantry Support

Wasp Carrier	Half-tracked	0	0	0	
Wasp flame-gun	*4"/10cm*	*3*	*-*	*6*	*Hull-mounted, Flame-thrower.*

Reconnaissance Vehicles

Universal Carrier	Half-tracked	0	0	0	Hull MG.
With Boys anti-tank rifle	*16"/40cm*	*2*	*4*	*5+*	*Hull mounted.*
With PIAT anti-tank projector	*8"/20cm*	*1*	*10*	*5+*	*Hull mounted.*
With .50 cal MG	*16"/40cm*	*3*	*4*	*5+*	*Hull mounted.*
With .extra MG	*16"/40cm*	*4*	*4*	*6*	*Hull mounted.*
Daimler Dingo	Jeep	1	0	0	AA MG.
Sawn Off Daimler	Jeep	1	0	0	AA MG.
Indian Pattern Carrier	Jeep	0	0	0	Hull MG.
With .50 cal MG	*16"/40cm*	*3*	*4*	*5+*	*Hull-mounted.*
Daimler I	Wheeled	1	0	0	Co-ax MG.
OQF 2 pdr gun	*24"/60cm*	*2*	*7*	*4+*	
With Littlejohn adaptor	*24"/60cm*	*2*	*9*	*5+*	*No HE.*
Humber LRC III	Jeep	0	0	0	MG.
Boys anti-tank rifle	*16"/40cm*	*2*	*4*	*5+*	*Hull mounted.*

Humber IV	Wheeled	1	0	0	Co-ax MG.
M6 37mm gun	*24"/60cm*	*2*	*7*	*4+*	
Daimler I	Wheeled	1	0	0	Co-ax MG.
OQF 2 pdr gun	*24"/60cm*	*2*	*7*	*4+*	
With Littlejohn adaptor	*24"/60cm*	*2*	*9*	*5+*	*No HE.*
Greyhound	Wheeled	1	0	0	Co-ax MG.
M6 37mm gun	*24"/60cm*	*2*	*7*	*4+*	
Staghound I	Wheeled	3	1	0	Co-ax MG, Hull MG.
M6 37mm gun	*24"/60cm*	*2*	*7*	*4+*	
Staghound II CS	Wheeled	3	1	0	Co-ax MG, AA MG.
OQF 3" howitzer	*24"/60cm*	*2*	*5*	*3+*	*Smoke.*
Firing bombardments	*40"/100cm*	-	*3*	*6*	*Smoke bombardment.*
Staghound AA	Wheeled	3	1	0	
Twin .50 cal Gun	*16"/40cm*	*5*	*4*	*5+*	*Anti-aircraft.*

Vehicle Machine-guns

Vehicle MG	*16"/40cm*	*3*	*2*	*6*	*ROF 1 if other weapons fire.*
.50 cal Vehicle MG	*16"/40cm*	*3*	*4*	*5+*	*ROF 1 if other weapons fire.*
Vehicle Twin MG	*16"/40cm*	*4*	*2*	*6*	*ROF 2 if other weapons fire.*

GUN TEAMS

Weapon	Mobility	Range	ROF	Anti-tank		Notes
Vickers HMG	Man-packed	24"/60cm	6	2	6	ROF 3 when pinned down or moving.
Firing bombardments		40"/100cm	-	-	-	
ML 3" Mk II mortar	Man-packed	24"/60cm	2	2	3+	Smoke, Minimum range 8"/20cm.
Firing bombardments		40"/100cm	-	2	6	Smoke bombardment.
ML 4.2" mortar	Light	48"/120cm	-	3	4+	Smoke bombardment.
Bofors 40mm gun	Immobile	24"/60cm	4	6	4+	Anti-aircraft, Gun shield, Turntable.
OQF 2 pdr gun	Medium	24"/60cm	3	7	4+	Gun shield, Turntable.
OQF 6 pdr gun	Medium	24"/60cm	3	10	4+	Gun shield.
OQF 17 pdr gun	Immobile	32"/80cm	2	13	3+	Gun shield, No HE.
OQF 3.7" gun	Immobile	40"/100cm	2	13	3+	Heavy Anti-aircraft, Turntable.
M1A1 75mm pack howitzer	Light	16"/40cm	2	6	3+	Smoke.
Firing bombardments		64"/160cm	-	3	6	Smoke bombardment.
OQF 25 pdr gun	Heavy	24"/60cm	2	9	3+	Gun shield, Smoke, Turntable.
Firing bombardments		80"/200cm	-	4	5+	Smoke bombardment.
BL 5.5" gun	Immobile	32"/80cm	1	13	1+	Bunker buster.
Firing bombardments		88"/220cm	-	5	2+	

INFANTRY TEAMS

Team	Range	ROF	Anti-tank	Firepower	Notes
Pistol team	4"/10cm	1	1	6	
Rifle team	16"/40cm	1	2	6	
Rifle/MG team	16"/40cm	2	2	6	
MG team	16"/40cm	3	2	6	ROF 2 when pinned down.
SMG team	4"/10cm	3	1	6	Full ROF when moving.
Light Mortar team	16"/40cm	1	1	4+	Smoke, Can fire over friendly teams.
PIAT team	8"/20cm	1	10	5+	Tank Assault 4.
Flame-thrower team	4"/10cm	2	-	6	Flame-thrower.
Staff team	16"/40cm	1	2	6	Moves as a Heavy Gun team.

Additional Training and Equipment

Pioneer teams are rated as Tank Assault 3.

Teams with Sticky Bombs are rated as Improvised Tank Assault 3.

Teams with Gammon Bombs are rated as Tank Assault 3.

TRANSPORT TEAMS

Vehicle	Mobility	Front	Side	Top	Equipment and Notes
Jeep or Jeep and Trailer	Jeep	-	-	-	
Recce Jeep	Jeep	-	-	-	Twin AA MG. *(Turning Tide)*
Recce Jeep	Jeep	-	-	-	Passenger-fired AA MG. *(A Bridge Too Far)*
CMP, Bedford, or Cut-down 15 cwt or 3-ton truck	Wheeled	-	-	-	
DUKW truck	Wheeled	-	-	-	Amphibious.
Handcart	Wagon	-	-	-	
White scout car or C15TA armoured truck	Jeep	1	0	0	
M5 half-track	Half-tracked	1	0	0	
Defrocked Priest APC	Standard Tank	1	0	0	.50 cal AA MG.
Troop, OP, Mortar, or Loyd Carrier	Half-tracked	0	0	0	
MMG Carrier	Half-tracked	0	0	0	HMG Carrier, Passenger-fired hull MG.
Indian Pattern Troop Carrier	Jeep	0	0	0	
Humber scout car	Jeep	1	0	0	AA MG.
Captured Sd Kfz 251/1 half-track	Half-tracked	1	0	0	Passenger-fired .50 cal AA MG.
Captured StuG	Standard Tank	7	3	1	Schürzen.
Tractors					
Quad, Matador, or Morris AA tractor	Wheeled	-	-	-	
Recovery Vehicles					
Sherman ARV	Standard Tank	6	4	1	Recovery vehicle.
Cromwell ARV	Light Tank	6	4	1	Recovery vehicle.
Churchill ARV	Slow Tank	8	7	2	Recovery vehicle, Wide tracks.

AIRCRAFT

Aircraft	Weapon	To Hit	Anti-tank	Firepower	Notes
Typhoon	Cannon	3+	8	5+	
	Rockets	3+	6	3+	
Hurricane IIC	Cannon	3+	8	5+	
	Bombs	4+	5	2+	
Kittyhawk	MG	3+	6	5+	
	Bombs	4+	5	1+	
Spitfire	MG	3+	7	5+	
	Bombs	4+	5	2+	

SOVIET

NATIONAL RULES

DOCTRINE

KOMISSAR

A Komissar team may attempt to force their troops to obey orders after failing a Motivation Test to Rally from being Pinned Down, to Counterattack in an assault (provided the Komissar is an Assaulting team), or to pass a Platoon Morale Check (but not for Tank Terror or any other Motivation Tests).

A Komissar attempting to force their troops to obey orders must be In Command, and must immediately Destroy one Infantry or Gun team from their own platoon (Soviet company) that is within the Command Distance of the Komissar team. The Komissar team cannot Destroy a Warrior team, Independent team, or itself in this way.

If you roll a 1 for a Motivation Test re-rolled using this rule, the soldiers are pushed past the point of breaking and revolt against the Komissar! In this case the Komissar team is Destroyed as well.

If the Company Command team of the company (Soviet battalion) or a Higher Command team has joined the company, re-roll the Motivation Test for the commander's influence first. If the unit fails the re-roll, the Komissar can then attempt to force the platoon (Soviet Company) to obey orders. This is an exception to the rule preventing a test from being re-rolled more than once (see page 20 of the rulebook).

BATTALION KOMISSAR

The Battalion Komissar team in the company (Soviet battalion) HQ is a Warrior team. If the Battalion Komissar team Joins a platoon (Soviet company), it takes over the duties of the normal Komissar team. The Komissar team has no effect as long as the Battalion Komissar team remains joined to the platoon (Soviet company).

DECOY

A Decoy Platoon must be identified before the enemy can shoot at it or assault it. These units start the game unidentified.

If a Decoy Platoon shoots or launches an assault, they are immediately identified by all enemy platoons.

In order to identify a Decoy Platoon, an enemy platoon must attempt to shoot at it or select it as the target for an artillery bombardment. Before rolling to hit with the shooting platoon, roll a Skill Test for the platoon.

If they pass the Skill Test, they identify the Soviet platoon or company and shoot at it as normal. The Soviet unit is now identified to all enemy platoons.

If they fail, they do not identify the unit, but may shoot at other Soviet units instead.

Aircraft cannot choose an unidentified team as the target of an air strike.

Enemy reconnaissance teams automatically identify a Decoy Platoon if they are within 8"/20cm when they attempt to shoot at them.

Sometimes your Decoy teams will be facing enemy other than the Germans. On these occasions the Decoy Platoon retains all the Decoy Special Rules as it is assumed that Soviets are using their new enemy's captured vehicles. However, you can still model your decoy troops as, and use the characteristics of, the historically-correct captured German equipment.

CENTRALISED CONTROL

Wherever the rules talk about a platoon, you can also read that as a Soviet company. Where the rules talk about a company, you can also read that as a Soviet battalion.

INFILTRATION

A platoon that has the Infiltration special rule lists the types of platoon (Soviet company) that they can Infiltrate with. The platoon may Infiltrate with one platoon (Soviet company) of the listed type, and any Warrior teams that Join them, when they are attacking in a mission with Prepared Positions and have the first turn.

Instead of making a Recce Deployment Move, the Infiltration platoon and the platoon and Warriors Infiltrating with it move up to 16"/40cm (regardless of their Movement Distance), remaining within Command Distance of each other, and taking any rolls needed to cross Rough Terrain and Obstacles. Their movement may not take them within 8"/20cm of any enemy team.

Any Infiltrating team that ends this movement within 16"/40cm and in of Line of Sight of an enemy team is automatically detected if they are either not Concealed from the enemy team, or the enemy team is a Recce team.

If the Infiltrating troops are not automatically detected, the opposing player rolls a die:

- On a roll of 6, they are detected.
- Otherwise, they have infiltrated undetected.

If the Infiltrators are detected, the Infiltrating player gives the first turn to their opponent.

Ignore enemy Warrior and Independent teams when Infiltrating. They do not limit an Infiltrating team's movement in any way and cannot detect them.

ARTILLERY

Steel Wall

Artillery batteries (Soviet artillery battalions) combining 76mm guns (such as the 76mm ZIS-3), 122mm howitzers (such as the 122mm obr 1938), and 120mm mortars (such as the 120-PM-38) may fire their 122mm howitzers and 120mm mortars as a separate battery apart from their 76mm guns.

If any 122mm howitzer or 120mm mortar fires an Artillery Bombardment, you must elect to either:

- include any or all of the 76mm guns in the Bombardment (using the Mixed Bombardment rule on page 131 of the rulebook), but not shoot with the remaining 76mm guns, or
- exclude all of the 76mm guns from the Artillery Bombardment, shooting normally with the 76mm guns instead.

This rule overrides the No Splitting of Artillery Fire rule on page 123 of the rulebook.

Roll Up the Guns

Soviet Medium and Heavy Gun teams and Staff teams move as if they were Light Gun teams.

Volley Fire

If they did not move in the Movement Step, Gun teams and Turretless Tank teams (such as the SU-76, SU-122, and ISU series) may re-roll failed rolls To Hit when shooting, provided the easiest team to hit in the target platoon is within 16"/40cm.

It takes time to build up the volume of fire that Soviet gunners prefer, making it impossible to use volley fire tactics effectively in a hastily-organised ambush. The Volley Fire rule does not apply to vehicle machine-guns, HMG teams or any Anti-aircraft weapon, nor when firing Artillery Bombardments, in Defensive Fire, or conducting an Ambush or Ambuscade.

INFANTRY

Quality of Quantity

A platoon (Soviet company) with at least fifteen Infantry or Gun teams In Command has Quality of Quantity. Include the Komissar team, any attached teams, and any Warrior and Independent teams that have Joined it when determining if a platoon (Soviet company) has Quality of Quantity.

A platoon (Soviet company) starting the Shooting Step with Quality of Quantity must take at least ten hits (instead of the usual five) in a single Shooting Step before they become Pinned Down, although a single hit from an Artillery Bombardment still Pins them Down.

Teams from a platoon (Soviet company) starting the Assault Step with Quality of Quantity may Charge into Contact and Counterattack from 12"/30cm away from the enemy rather than 8"/20cm as usual. Teams further than 12"/30cm are still Non-assaulting teams.

A platoon (Soviet company) with at least fifteen Assaulting Infantry teams must take at least ten hits in Defensive Fire before they are Pinned Down and forced to Fall Back.

Shtraf Companies

Any battalion fielding a Shtraf Company Always Attacks in any mission. As an assault unit, a Shtraf Company can never attempt to Dig In and never benefits from the Prepared Positions special rule.

Soldiers were sent to serve in a Shtraf Company because of failure on the battlefield. As such, a Shtraf Company always passes a Motivation Test on a roll of 2+.

Because it is expected to die, a Shtraf Company is totally ignored when taking a Company Morale Check. It does not add to the number of operational platoons, nor does it count as Destroyed if it suffers that fate.

Hen and Chicks

If a Platoon Command team moves in the Movement Step, all of the Tank and Transport teams in its platoon (Soviet company) must move as well.

If the Platoon Command team does not move, only Tank and Transport teams that started the Movement Step Out of Command may move.

Any vehicle that moves adds +1 to the score required to hit when shooting with its main gun or guns. This gives a cumulative penalty of +2 for ROF 1 weapons. Machine-guns and Flame-throwers do not suffer this penalty.

Cupolas

Tanks fitted with a commander's cupola do not suffer the effects of Limited Vision (see page 118 of the rulebook).

Turret-Rear Machine-gun

A Turret-rear MG may not shoot, it is purely defensive.

In assault combat, if a hit from an Infantry or Gun team is allocated to a Tank team armed with a Turret-rear MG, the team that scored the hit must take another Skill Test.

- If they pass the second Skill Test, the assaulting team gets past the machine-gun to hit the tank and the tank must take an Armour Save as normal.
- If they fail the second Skill Test, the machine-gunner managed to keep the infantry at bay and protect the tank, and the assaulting team fails to score a hit.

Bailed out Tank teams do not benefit from their Turret-rear MG.

Limited Vision

Tanks with Limited Vision add +1 to the score required to hit when shooting any weapon except an AA MG at targets that are entirely behind a line drawn across the front of the tank's turret before they rotate their turret to face the target.

Tanks with Limited Vision must rotate their turret to point straight ahead when they move.

Tankodesantniki

Tanks with Tankodesantniki have Tank Escorts (see pages 200 and 201 of the rulebook).

Tank-rider Companies

Teams from Tank-rider Companies have a 3+ Save while Mounted on a Tank team as Passengers. They do not need to Dismount Under Fire when hit. They can remain on the tanks instead.

PULYEMYETNO-ARTILLYERIYSKIY BATALON

Camouflaged Positions

An enemy team must be able to draw a Line of Sight to and be within 16"/40cm of a team from a Pulyemyetno-Artillyeriyskiy Batalon HQ, Pulyemyetno-Artillyeriyskiy Company and Artillery Group that are in Trench Lines or Gun Pits and has not shot during the game to be able to shoot at them.

No Retreat, No Surrender

A Pulyemyetno-Artillyeriyskiy Company does not use the Quality of Quantity rule, instead they may re-roll failed Motivation tests to Counterattack in assaults.

8YA OTDYELNAYA SHTRAFNOY BATALON

Assault Force

The 8ya Otdyelnaya Shtrafnoy Batalon Always Attacks in any mission.

Since it is expected to always go forward, Companies in the 8ya Otdyelnaya Shtrafnoy Batalon can never attempt to Dig In and never benefit from the Prepared Positions special rule.

Initiative is Punishable

The 8ya Otdyelnaya Shtrafnoy Batalon never has to take a Company Morale Check for being below half strength. The Companies still take Platoon Morale Checks as normal, though. However, if no Shtraf Companies (Combat Companies) remain on the table, the remaining forces automatically fail a Company Morale Check and the game ends.

Shtrafnik's Absolution

8ya Otdyelnáya Shtrafnoy Batalon HQ, Shtraf Companies, Shtraf Machine-gun Companies and Shtraf Mortar Companies always pass a Motivation Test on a roll of 2+.

UDARNY STRELKOVY BATALON

Guards Assault Battalion

An Udarny Strelkovy Batalon Always Attacks against any other Infantry Company.

Hardened Veterans

Infantry teams from a Udarny Strelkovy Batalon HQ, Udarny Strelkovy Company, or Shock Submachine-gun Company hit on a roll of 3+ in assault combat.

COSSACKS

Urah Pobeda!

See the Cavalry Charge on Through rule on page 203 of the rulebook.

Tachanka

The Tachanka is an HMG Carrier (see page 118 of the rulebook).

Mounted Man-packed Guns

See pages 202 to 204 of the rulebook.

PARTIZANSKIY BRIGADA

Breakthrough Forces

In missions that have Reserves of any type (Reserves, Scattered Reserves, Delayed Reserves, etc.) Corps Support Companies in a Partizanskiy Brigada must be held in Reserve.

Our Land

Infantry and Man-packed Gun teams from Partizanskiy Brigada HQ, Combat and Weapons Companies of a Partizanskiy Brigada treat Difficult Going as Cross-country terrain, allowing them to move at the double.

Partisan Ambush

When defending in a Mission with the Prepared Positions special rule a Partizanskiy Brigada may deploy a Partizanskiy Company in Immediate Ambush (see page 267 of the rulebook) in addition to any Ambush deployment that may be allowed by the Mission played. The Immediate Ambushing platoon must be taken from the platoons that are to be deployed on table during deployment.

SHTURMOVYE AND SAPPER

Smoke Pots

A platoon (Soviet company) with Smoke Pots may designate Rifle, Rifle/MG, MG or SMG teams as Smoke Pot teams to lay smoke for an assault. A Smoke Pot team cannot shoot in the Shooting Step nor assault in the Assault Step, but acts as if giving Covering Fire during the Assault Step.

When the platoon (Soviet company) launches its assault, roll a Skill Test for each Smoke Pot team within 4"/10cm of a platoon being assaulted. If any Smoke Pot team passes the Skill Test then all teams from that platoon (Soviet company) participating in the assault are Concealed.

Sapper Body Armour

A team with Sapper Body Armour rolls a die when hit during an Assault.

If they score a 6, they are unharmed, but still count as hit for the purposes of requiring a Motivation Test to Carry on with the Assault or Fight Back.

If they roll any other score, they are Destroyed and removed from the table as usual.

Faustnika

Faustnika teams that move in the Movement Step can only shoot in the Shooting Step if they are shooting at an enemy Bunker or an Infantry or Gun team in a Building.

Faustnika teams are Bunker Busters when shooting at an enemy Bunker or an Infantry or Gun team in a Building.

EXPLOITATION FORCE

Exploitation Force

A Gvardeyskiy Tankovy Batalon, Peredovoye Otryad, Rota Razvedki, Gvardeyskiy Kazachiy Polk, and Inomarochnikiy Tankovy Batalon are Exploitation Forces. An Exploitation Force will Always Attacks.

LEYTENANT A V PYL'CYN

Leytenant A V Pyl'cyn is a Warrior SMG team that is the Command team of one of the Shtraf Companies in your 8ya Otdyelnaya Shtrafnoy Batalon.

Reconnaissance in Force

Deploy Pyl'cyn's Shtraf Company when you would normally deploy Independent Teams.

Mine Expert

Pyl'cyn may gap Minefields and Barbed Wire Entanglements as though he was a Pioneer team.

Rita Makarievskaya

Model Rita Makarievskaya on the same base as Pyl'cyn.

Once per enemy Shooting Step, Rita may attempt to save a team that is in Pyl'cyn's Shtraf Company. If an Infantry or Gun team within 6"/15cm of Pyl'cyn fails a Save, roll a die:

- If the result is 5+, Rita manages to save the team and it continues to fight on unharmed.
- Otherwise, the soldiers are too badly wounded and the team is Destroyed as normal.

KAPITAN V I NEVSKY

Kapitan V I Nevsky is a Warrior Company Command Tank team rated as Fearless Veteran.

Nevsky may join any Inomarochnikiy Tankovy Batalon and replaces the Company Command M4 (M4A2 Sherman) or M4 76mm (M4A2 Sherman) tank in the Battalion HQ for an additional +50 points.

Served on the Staff

After all deployment, but before the game begins, Kapitan V I Nevsky may move himself and one Combat Company within 6"/15cm of him, that is not already part of an Infiltration, up to their normal distance in any direction within their deployment area. This movement may not be At the Double and may not take any team outside the deployment area.

Respect of the Men

Kapitan V I Nevsky's tank and any platoon (Soviet company) he is currently leading always pass Motivation Tests on a roll of 2+.

Battle Tested

The Hen and Chicks special rule does not apply to Nevsky. If he joins a platoon (Soviet company), the rest of the platoon still obey the Hen and Chicks special rule.

BRIGADA KOMISSAR M A DEDOV

Add Brigada Komissar M A Dedov to a Gvardeyskiy Tankovy Company from a Gvardeyskiy Tankovy Batalon or Tankovy Company from a Tankovy Batalon for +70 points. M A Dedov is rated Fearless Veteran.

Brigada Komissar M A Dedov is a Warrior. He takes over a tank from a Gvardeyskiy Tankovy Company or Tankovy Company at the beginning of the game before deployment. M A Dedov and his tank are no longer a part of any platoon (Soviet company).

YOU OVER THERE, WHAT ARE YOU DOING TO DEFEAT THE FASCISTS?

Brigada Komissar M A Dedov acts as a 2iC Command team and follows all the rules for 2iC Command teams on pages 105 and 175 in the rulebook.

COMRADES FORWARD!

Any Soviet company entirely equipped with Tank teams joined by Brigada Komissar M A Dedov may re-roll any Motivation Tests it and its members are required to make.

When leading a platoon (Soviet company) with other types of teams in it, Brigada Komissar M A Dedov uses the Komissar special rules on page 249 of the rulebook. However, if you roll a 1 for the new Motivation Test the platoon (Soviet Company) does not destroy Brigada Komissar M A Dedov, he instead becomes Bailed Out as he retreats back into his tank turret away from the angry soldiers.

PODPOLKOVNIK PAVEL KAMNEV

Pavel Kamnev is a Warrior rated as a Fearless Veteran. He is a mounted Company Command SMG team. He replaces the Company Command team of a Gvardeyskiy Kazachiy Polk for +50 points.

FEARLESS CHARGE

Pavel Kamnev and any Mounted Gvardeyskiy Kazachiy Company that he has joined make saves against hits from Defensive Fire on a 5+ instead of the usual 3+.

If the company is Pinned Down by Defensive Fire, he may take an immediate Motivation Test. If he passes the Company does not fall back from the Defensive Fire and continues the Assault.

MASTER SWORDSMAN

Pavel Kamnev rerolls failed Skill Tests to hit in assault combat.

ALWAYS SADDLED

Pavel Kamnev can never send his horse to the rear. If a platoon that he has joined sends their horses to the rear he will remain mounted.

If Kamnev uses the Warrior Infantry Team Casualties rule to avoid destruction he remains mounted, taking his aide's horse and sending the aide in search of a fresh mount.

LEYTENANT V S RAT

Leytenant Rat is a Warrior team who leads a Guards Heavy Assault Gun Company. He takes over the Command Assault gun for +45 points. Leytenant Rat and his assault gun are rated Fearless Trained.

KEEP'EM GUESSING

Leytenant Rat's assault gun does not suffer the normal +1 To Hit penalty for shooting while moving with a ROF 1 gun.

CRAZY VASILY 'ZIG-ZAG'

All ISU-122 and ISU-152 assault guns in Rat's company, that are not Bogged Down or Bailed Out, who moved at least 6"/15cm towards any enemy team or Objective in their previous Movement Step, and are hit in the enemy Shooting Step, may roll a die for each hit they are allocated.

- On a score of 5+, Rat's zig-zag is successful and the hit is ignored.
- Otherwise the hit is resolved as normal.

LEYTENANT IVAN ZABOLOTNY

Leytenant Ivan Zabolotny is a Warrior Recce Pioneer SMG team. He replaces the Command Pioneer SMG team in an Inzhenerno-Saperniy Company for +70 points. Zabolotny is rated as Fearless Trained.

FOR BRAVERY!

Zabolotny and his Inzhenerno-Saperniy Company always pass Motivation Tests on a 2+.

LEADING THE WAY!

If your force does not contain another platoon with the Infiltration special rule you may Infiltrate with Zabolotny.

Zabolotny may Infiltrate his Inzhenerno-Saperniy Company. See the Infiltration rules on page 250 of the rulebook.

GVARDEYSKIY KAPITAN DMITRIY LOZA

Gvardeyskiy Kapitan Dmitriy Loza is a Warrior Command Tank team rated as Fearless Veteran.

Loza may join any Perdovoye Otryad or Inomarochnikiy Tankovy Batalon and replaces the Command M4 or M4 76mm tank in an Inomarochnikiy Tankovy Company for an additional +80 points.

SECOND IN COMMAND

Gvardeyskiy Kapitan Dmitriy Loza may take Company Morale Checks as if he were the Company Command team if the Company Command team is Destroyed.

SPETSNAZ LOZA

Loza is an Infiltration Platoon. Loza may Infiltrate his Inomarochnikiy Tankovy Company.

BATTLE TESTED

The Hen and Chicks special rule does not apply to Loza. The rest of his Inomarochnikiy Tankovy Company still obey the Hen and Chicks special rule.

EMCHA GUNNER

The M3 75mm or M1 76mm gun of Loza's M4 or M4 76mm tank has ROF 3 when stationary and ROF 2 when moving.

SOVIET ARSENAL

TANK TEAMS

Name *Weapon*	Mobility *Range*	Armour Front *ROF*	Armour Side *Anti-tank*	Armour Top *Firepower*	Equipment and Notes
LIGHT TANKS					
T-70 obr 1943	Standard Tank	4	2	1	Co-ax MG, Limited vision, Wide-tracks.
45mm obr 1938 gun	*24"/60cm*	*1*	*7*	*4+*	
Mark III (Valentine II)	Slow Tank	6	5	1	Co-ax MG.
OQF 2 pdr gun	*24"/60cm*	*2*	*7*	*4+*	
Mark III (Valentine VIII)	Slow Tank	6	4	1	Protected ammo.
OQF 6 pdr gun	*24"/60cm*	*2*	*10*	*4+*	
M3l (M3A1 Stuart)	Light Tank	3	2	1	Co-ax MG, Hull MG.
M6 37mm gun	*24"/60cm*	*2*	*7*	*4+*	
MEDIUM TANKS					
T-34 obr 1942	Standard Tank	6	5	1	Co-ax MG, Hull MG, Fast tank, Limited vision, Wide-tracks.
76mm F-34 gun	*32"/80cm*	*2*	*9*	*3+*	
T-34/85 obr 1943	Standard Tank	7	5	1	Co-ax MG, Hull MG, Limited vision.
85mm D-5T gun	*32"/80cm*	*2*	*12*	*3+*	
T-34/85 obr 1944	Standard Tank	7	5	1	Co-ax MG, Hull MG.
85mm ZIS-53 gun	*32"/80cm*	*2*	*12*	*3+*	
Mark II (Matilda II)	Slow Tank	7	6	2	Co-ax MG, Unreliable.
OQF 2 pdr gun	*24"/60cm*	*3*	*7*	*4+*	
Mark II 76mm (Matilda II CS)	Slow Tank	7	6	2	Co-ax MG, Unreliable.
OQF 3" gun	*24"/60cm*	*2*	*5*	*3+*	
M3s (M3 Lee)	Standard Tank	5	3	1	Co-ax MG, Cupola MG.
M6 37mm gun	*24"/60cm*	*3*	*7*	*4+*	
M3 75mm gun	*32"/80cm*	*2*	*10*	*3+*	*Hull mounted.*
M4 (M4A2 Sherman)	Standard Tank	6	4	1	Co-ax MG, Hull MG.
M3 75mm gun	*32"/80cm*	*2*	*10*	*3+*	
M4 76mm (M4A2 Sherman)	Standard Tank	7	4	1	Co-ax MG, Hull MG, Protected Ammo.
M1 76mm gun	*32"/80cm*	*2*	*12*	*3+*	
HEAVY TANKS					
KV-1e	Slow Tank	9	8	2	Co-ax MG, Hull MG, Turret-rear MG, Limited vision, Unreliable.
76mm F-34 gun	*32"/80cm*	*2*	*9*	*3+*	
KV-1s	Standard Tank	8	6	2	Co-ax MG, Hull MG, Turret-rear MG, Wide tracks.
76mm F-34 gun	*32"/80cm*	*2*	*9*	*3+*	
KV-85	Slow Tank	9	7	2	Co-ax MG, Turret-rear MG.
85mm D-5T gun	*32"/80cm*	*2*	*12*	*3+*	
IS-85	Slow Tank	10	8	2	Co-ax MG, Turret-rear MG.
85mm D-5T gun	*32"/80cm*	*2*	*12*	*3+*	
IS-2 obr 1943	Slow Tank	10	8	2	Co-ax MG, Turret-rear MG.
122mm D-25T gun	*32"/80cm*	*1*	*15*	*2+*	*Breakthrough gun.*
Mark IV (Churchill III or IV)	Slow Tank	8	7	2	Co-ax MG, Hull MG, Protected ammo, Wide tracks.
OQF 6 pdr gun	*24"/60cm*	*3*	*10*	*4+*	
ASSAULT GUNS					
SU-57	Half-tracked	1	0	0	
M1 57mm gun	*24"/60cm*	*3*	*10*	*4+*	*Hull mounted.*
SU-76M	Standard Tank	3	1	0	Wide-tracks.
76mm ZIS-3 gun	*32"/80cm*	*2*	*9*	*3+*	*Hull mounted.*
M-10	Standard Tank	4	2	0	.50 cal AA MG.
M7 3in gun	*32"/80cm*	*2*	*12*	*3+*	*Slow traverse.*

SU-85	Standard Tank	5	5	1	
85mm D-5S gun	*32"/80cm*	*2*	*12*	*3+*	*Hull mounted.*
SU-85M	Slow Tank	9	5	1	Overloaded.
85mm D-5S gun	*32"/80cm*	*2*	*12*	*3+*	*Hull mounted.*
SU-100	Slow Tank	9	5	1	Overloaded.
100mm D-10S gun	*40"/100cm*	*1*	*16*	*2+*	*Hull mounted.*
SU-122	Standard Tank	5	5	1	
122mm obr 1938 howitzer	*24"/60cm*	*2*	*10*	*2+*	*Hull mounted, Breakthrough gun.*
HEAVY ASSAULT GUNS					
SU-152	Slow Tank	7	6	2	
152mm ML-20S gun	*32"/80cm*	*1*	*13*	*1+*	*Hull mounted, Bunker buster.*
ISU-122	Slow Tank	9	7	2	
122mm D-25S gun	*32"/80cm*	*1*	*15*	*2+*	*Hull mounted, Breakthrough gun.*
ISU-152	Slow Tank	9	7	2	
152mm ML-20S gun	*32"/80cm*	*1*	*13*	*1+*	*Hull mounted, Bunker buster.*
ENGINEERING TANKS					
OT-34	Standard Tank	6	5	1	Co-ax MG, Fast tank, Limited vision, Wide-tracks.
76mm F-34 gun	*32"/80cm*	*1*	*9*	*3+*	
ATO-42 flame-thrower	*4"/10cm*	*2*	*-*	*6*	*Hull mounted, Flame-thrower.*
OT-34/85 obr 1943	Standard Tank	7	5	1	Co-ax MG, Limited vision.
85mm D-5T gun	*32"/80cm*	*1*	*12*	*3+*	
ATO-42 flame-thrower	*4"/10cm*	*2*	*-*	*6*	*Hull mounted, Flame-thrower.*
KV-8s	Standard Tank	8	6	2	Hull MG, Turret-rear MG, Wide-tracks.
ATO-42 flame-thrower	*4"/10cm*	*3*	*-*	*6*	*Flame-thrower.*
45mm obr 1938 gun	*24"/60cm*	*2*	*7*	*4+*	
PT-34	Standard Tank	7	5	1	Co-ax MG, Mine roller, Limited vision.
76mm F-34 gun	*32"/80cm*	*2*	*9*	*3+*	
CAPTURED GERMAN TANKS					
Panzer IV H	Standard Tank	6	3	1	Co-ax MG, Hull MG, Protected ammo, Schürzen, Unreliable.
7.5cm KwK40 gun	*32"/80cm*	*1*	*11*	*3+*	
StuG III G	Standard Tank	7	3	1	Hull MG, Protected ammo, Schürzen, Unreliable.
7.5cm StuK40 gun	*32"/80cm*	*1*	*11*	*3+*	*Hull mounted.*
Tiger I E	Slow Tank	9	8	2	Co-ax MG, Hull MG, Protected ammo, Wide tracks, Unreliable.
8.8cm KwK36 gun	*40"/100cm*	*1*	*13*	*3+*	*Slow traverse.*
Panther D or A	Standard Tank	10	5	1	Co-ax MG, Hull MG, Wide tracks, Unreliable.
7.5cm KwK42 gun	*32"/80cm*	*1*	*14*	*3+*	

Armoured Cars

BA-64	Jeep	0	0	0	AA MG.
With PTRD anti-tank rifle	*16"/40cm*	*2*	*5*	*5+*	*Hull mounted.*

Rocket Launchers

BM-13-16 Katyusha	Wheeled	-	-	-	
BM-13-16 rocket launcher	*64"/160cm*	-	*2*	*4+*	*Rocket launcher.*
BM-31-12 Katyusha	Wheeled	-	-	-	
BM-31-12 rocket launcher	*48"/120cm*	-	*3*	*1+*	*Rocket launcher.*

Anti-aircraft Vehicles

DShK AA truck	Wheeled	-	-	-	
DShK AA gun	*16"/40cm*	*4*	*4*	*5+*	*Anti-aircraft, Portee.*
ZSU M17 MGMC (quad .50 cal)	Half-tracked	1	0	0	
M45 quad .50 cal gun	*16"/40cm*	*6*	*4*	*5+*	*Anti-aircraft.*

Cavalry Machine-guns

Tachanka	Wagon	-	-	-	
Maxim HMG	*24"/60cm*	*6*	*2*	*6*	*HMG Carrier.*

Vehicle Machine-guns

Vehicle MG	*16"/40cm*	*3*	*2*	*6*	*ROF 1 if other weapons fire.*
.50 cal Vehicle MG	*16"/40cm*	*3*	*4*	*5+*	*ROF 1 if other weapons fire.*

GUN TEAMS

Weapon	Mobility	Range	ROF	Anti-tank	Firepower	Notes
PTRD anti-tank rifle	Man-packed	16"/40cm	2	5	5+	Tank Assault 3.
Maksim HMG	Man-packed	24"/60cm	6	2	6	ROF 3 when pinned down or moving
82-BM-41 mortar	Man-packed	24"/60cm	2	2	3+	Minumum range 8"/20cm.
Firing bombardments		40"/100cm	-	2	6	
120-PM-38 mortar	Light	56"/140cm	-	3	3+	
76mm obr 1927 gun	Light	16"/40cm	2	5	3+	Gun shield.
Firing bombardments		64"/160cm	-	3	6	
122mm obr 1938 assault howitzer	Heavy	32"/80cm	1	7	2+	Bunker buster, Gun shield.
DShK AA MG	Man-packed	16"/40cm	4	4	5+	Anti-aircraft, Turntable.
37mm obr 1939 gun	Immobile	24"/60cm	4	6	4+	Anti-aircraft, Turntable.
45mm obr 1937 gun	Light	24"/60cm	3	7	4+	Gun shield.
45mm obr 1942 gun	Light	24"/60cm	3	8	4+	Gun shield.
57mm ZIS-2 gun	Heavy	32"/80cm	3	11	4+	Gun shield.
76mm ZIS-3 anti-tank gun	Heavy	32"/80cm	2	9	3+	Gun shield.
85mm obr 1939 gun	Immobile	32"/80cm	2	12	3+	Heavy anti-aircraft, Turntable.
100mm BS-3 gun	Immobile	40"/100cm	1	16	2+	Gun shield.
76mm ZIS-3 field gun	Heavy	32"/80cm	2	9	3+	Gun shield.
Firing bombardments		80"/200cm	-	3	6	
122mm obr 1938 howitzer	Immobile	24"/60cm	1	7	2+	Breakthrough gun, Gun shield.
Firing bombardments		80"/200cm	-	4	3+	
122mm A-19 obr 1931/37 gun	Immobile	32"/80cm	1	15	2+	Breakthrough gun, Gun shield.
Firing bombardments		96"/240cm	-	4	3+	
152mm obr 1943 howitzer	Immobile	24"/60cm	1	10	1+	Bunker buster, Gun shield.
Firing bombardments		80"/200cm	-	5	2+	
152mm ML-20 obr 1931/37 howitzer	Immobile	32"/80cm	1	13	1+	Bunker buster, Gun shield.
Firing bombardments		88"/220cm	-	5	2+	

INFANTRY TEAMS

Team	Range	ROF	Anti-tank	Firepower	Weapons and Notes
Rifle team	16"/40cm	1	2	6	
Rifle/MG team	16"/40cm	2	2	6	
MG team	16"/40cm	3	2	6	ROF 2 when pinned down
SMG team	4"/10cm	3	1	6	Full ROF when moving.
Light Mortar team	16"/40cm	1	1	4+	Can fire over friendly teams.
Faustnika team	4"/10cm	2	12	5+	Tank Assault 6, Cannot shoot in Shooting Step if moved in Movement Step, Faustnika, Bunker buster.
Flame-thrower team	4"/10cm	4	-	6	Flame-thrower.
Komissar team	4"/10cm	1	1	6	
Staff team	16"/40cm	1	2	6	Moves as a Light Gun team.

Additional Training and Equipment

Pioneer teams are Tank Assault 4.

TRANSPORT TEAMS

Vehicle	Mobility	Front	Side	Top	Equipment and Notes
Trucks					
Ford GPA amphibious jeep	Jeep	-	-	-	Amphibious
ZIS-5 3-ton, ZIS-6 4-ton, Dodge ½-ton, or Studebekker 2½-ton truck	Wheeled	-	-	-	
Horse-drawn wagon	Wagon	-	-	-	
Tractors					
Stalinets	Slow Tank	-	-	-	
Horse-drawn limber	Wagon	-	-	-	
Armoured Transporters					
M3A1 armoured transporter	Jeep	1	0	0	Passenger-fired AA MG or .50 cal AA MG.
M2 or M3 half-tracks	Half-tracked	1	0	0	Optional Passenger-fired .50 cal AA MG.
Turretless BA-10M armoured transporter	Jeep	1	0	0	Passenger-fired MG.
Captured Sd Kfz 250 or Sd Kfz 251 half-track	Half-tracked	1	0	0	Passenger-fired MG.
Universal Carrier	Half-tracked	0	0	0	Passenger-fired MG.

AIRCRAFT

Aircraft	Weapon	To Hit	Anti-tank	Firepower	Notes
Il-2 Shturmovik	Cannon	3+	9	5+	Flying tank.
	Bombs	4+	5	1+	
	Rockets	3+	6	3+	
Il-2M Tip 3M Shturmovik	Cannon	3+	12	4+	Flying tank.
	Bombs	4+	5	1+	
	Rockets	3+	6	3+	

FORTIFICATIONS

Weapon	Mobility	Range	ROF	Anti-tank	Firepower	Notes
Maksim HMG Nest	Immobile	24"/60cm	6	2	6	ROF 3 when pinned down.
45mm obr 1942 Nest	Immobile	24"/60cm	3	8	4+	

MINOR NATIONS

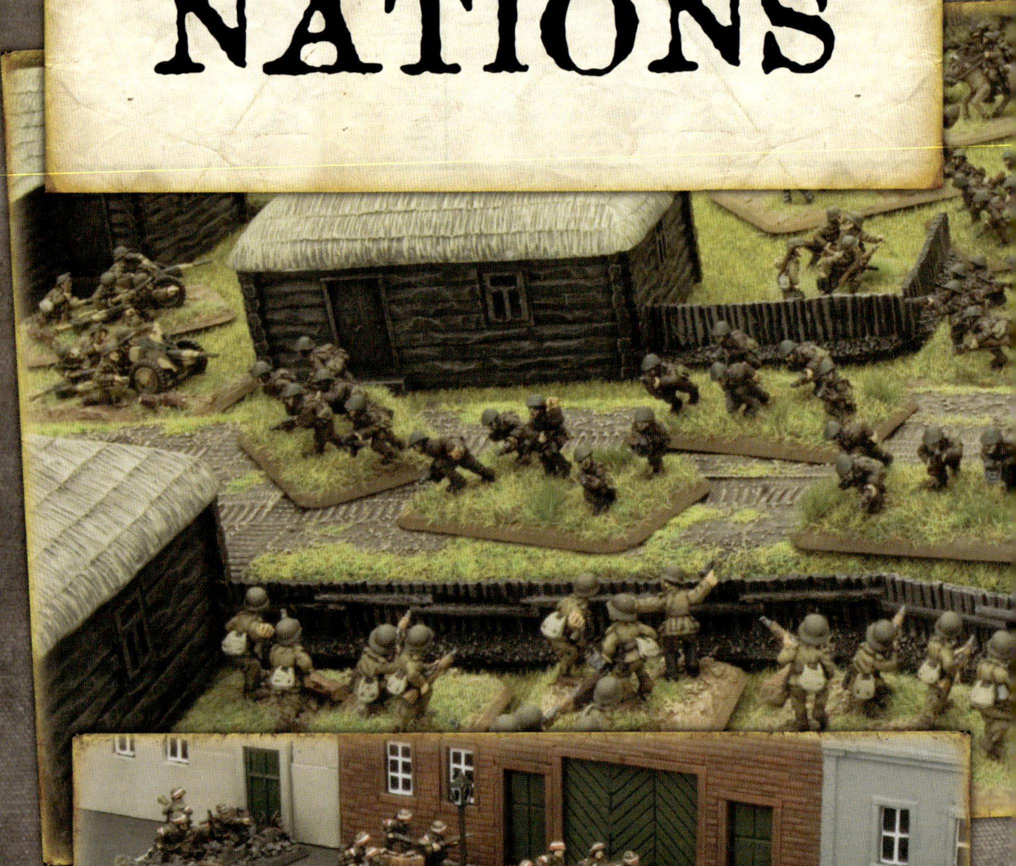

FINNISH

SELF SUFFICIENT

Finnish Platoons use the German Mission Tactics special rule (see page 242 of the rulebook).

HUNTERS

Finnish Infantry and Man-packed Gun teams may move At the Double through Woods and deep snow.

BITTER ENEMIES

When Finnish platoons fighting against any Soviet force take a Motivation Test to Counterattack in assaults, you may re-roll the die and apply the re-rolled result to all Finnish platoons that failed on the original roll.

INTERCEPTED COMMUNICATIONS

Because of the excellent quality of the Finnish radio interception, Artillery Bombardments fired by an artillery platoon with a Staff team may re-roll the third failed attempt to range in. The artillery platoon does not get a re-roll on their first and second attempts.

Note: Finnish forces from the Lapin Sota PDF Intelligence Briefing do not use the Intercepted Communications special rule.

AUTOMATED FIRE CONTROL

A Finnish Observer team from an artillery battery with a Staff team may act as the Spotting team for any artillery platoon, whether or not they have a Staff team.

When a Finnish artillery battery that has a Staff team fires an Artillery Bombardment, any other artillery batteries with Staff teams may join it before rolling to Range In and fire as a single combined Artillery Battery using the Mixed Bombardments rule on page 131 of the rulebook.

KORPRAALI TOIVO ILOMÄKI

Toivo Ilomäki is a Warrior who may be added to upgrade one anti-tank gun team of a Heavy Anti-Tank Platoon for +10 points. He is a Fearless Veteran team armed with the same anti-tank gun as the rest of the platoon.

NERVES OF STEEL

Toivo Ilomäki's Gun team is never Pinned Down, regardless of the platoon's status.

ONE SHOT, ONE KILL

Toivo Ilomäki's Gun team may re-roll any failed To Hit rolls.

WARRIOR GUN TEAM

Toivo Ilomäki is a Warrior Gun team.

KAPTEENI LAURI TÖRNI

Captain Lauri Törni is a Warrior Command SMG team rated as Fearless Veteran. Törni replaces the Command team of a Jääkäri Platoon for a cost of +65 points. Lauri counts as a Recce team.

He is armed with a Suomi SMG and Satchel Charges. The Satchel Charges gives Captain Lauri Törni's Warrior Command SMG team Tank Assault 4 rating.

Implacable

Törni and the Jääkäri Platoon he has joined pass all Motivation Tests on a 2+.

Skilled Soldier

Törni and the Jääkäri Platoon he has joined may re-roll any failed Skill Tests.

COLONEL ADOLF EHRNROOTH

Adolf Ehrnrooth is a Warrior Higher Command Pistol team who may be added to an Elite Jalkaväki Komppania for +40 points. He is a Fearless Veteran armed with a pistol.

Inspiring Leader

A Komppania including Ehrnrooth may re-roll failed Company Morale Checks.

Mannerheim's Men

Any platoon Ehrnrooth has joined passes Motivation on a 2+.

FINNISH ARSENAL

TANK TEAMS

Name *Weapon*	Mobility *Range*	Armour Front *ROF*	Side *Anti-tank*	Top *Firepower*	Equipment and Notes
TANKS					
T-26	Slow Tank	1	1	1	Co-ax MG, Limited vision, Unreliable.
45mm obr 1934 gun	*24"/60cm*	*2*	*7*	*4+*	
T-28	Slow Tank	4	3	2	Two Deck-turret MG, Turret-front MG, Turret-rear MG, Limited vision, Unreliable.
76mm L-10 gun	*24"/60cm*	*2*	*7*	*3+*	
T-34/76	Standard Tank	6	5	1	Co-ax MG, Hull MG, Fast tank, Limited vision, Wide track.
76mm F-34 gun	*32"/80cm*	*2*	*9*	*3+*	
T-34/85	Standard Tank	7	5	1	Co-ax MG, Hull MG, Limited vision.
85mm D-5T gun	*32"/80cm*	*2*	*12*	*3+*	
KV-1e	Slow Tank	9	8	2	Co-ax MG, Hull MG, Turret-rear MG, Limited vision, Unreliable.
76mm F-34 gun	*32"/80cm*	*2*	*9*	*3+*	
ASSAULT GUNS					
Stu 40 G Sturmi	Standard Tank	7	3	1	Hull MG, Protected ammo, Improvised Armour.
7.5cm StuK40 gun	*32"/80cm*	*2*	*11*	*3+*	*Hull mounted.*
BT-42	Standard Tank	1	1	1	Limited vision.
114 Psv.H/18 howitzer	*16"/40cm*	*1*	*7*	*2+*	*Breakthrough gun.*
Firing bombardments	*48"/120cm*	*-*	*4*	*4+*	
ISU-152	Slow Tank	9	7	2	
152mm ML-20S gun	*32"/80cm*	*1*	*13*	*1+*	*Hull mounted, Bunker buster.*
SELF-PROPELLED ANTI-AIRCRAFT					
Landsverk Anti II	Standard Tank	2	1	0	
40 ItK/38 gun	*24"/60cm*	*4*	*6*	*4+*	*Anti-aircraft.*
ARMOURED CARS					
BA-10	Wheeled	1	0	0	Co-ax MG, Hull MG, Limited vision.
45mm obr 1934 gun	*24"/60cm*	*2*	*7*	*4+*	
VEHICLE MACHINE-GUNS					
Vehicle MG	*16"/40cm*	*3*	*2*	*6*	*ROF 1 if other weapons fire.*

GUN TEAMS

Weapon	Mobility	Range	ROF	Anti-tank	Firepower	Notes
Maxim HMG	Man-packed	24"/60cm	6	2	6	ROF 3 when pinned down or moving.
Lahti anti-tank rifle	Man-packed	16"/40cm	3	5	5+	
Tampella M/35 81mm mortar	Man-packed	24"/60cm	2	2	3+	Smoke, Minimum range 8"/20cm.
Firing bombardments		40"/100cm	-	2	6	Smoke bombardment.
Tampella M/40 120mm mortar	Light	56"/140cm	-	3	3+	Smoke bombardment.
37 PstK/37 gun	Light	24"/60cm	3	6	4+	Gun shield.
Firing Stielgranate		8"/20cm	1	12	5+	
45 PstK/37 gun	Light	24"/60cm	3	7	4+	Gun shield.
50 PstK/38 gun	Medium	24"/60cm	3	9	4+	Gun shield.
75 PstK/97-38 gun	Medium	24"/60cm	2	10	3+	Gun shield.
75 PstK/40 gun	Medium	32"/80cm	2	12	3+	Gun shield.
20 LtK/38 gun	Light	16"/40cm	4	5	5+	Anti-aircraft, Turntable.
40 LtK/38 gun	Immobile	24"/60cm	4	6	4+	Anti-aircraft, Turntable.

76 RK/27 gun	Light	16"/40cm	2	5	3+	Gun shield.
Firing bombardments		64"/160cm	-	3	6	
76 K/02 gun	Heavy	24"/60cm	2	8	3+	Gun shield, Smoke.
Firing bombardments		60"/160cm	-	3	6	Smoke bombardment.
105 H/33 howitzer	Immobile	24"/60cm	1	10	2+	Gun shield, Breakthrough gun, Smoke.
Firing bombardments		72"/180cm	-	4	4+	Smoke bombardment.
122 H/38 howitzer	Immobile	24"/60cm	1	7	2+	Gun shield, Breakthrough gun.
Firing bombardments		80"/200cm	-	4	3+	
152 H/43 howitzer	Immobile	24"/60cm	1	10	1+	Bunker buster, Gun shield.
Firing bombardments		80"/200cm	-	5	2+	
150 H/40 howitzer	Immobile	24"/60cm	1	13	1+	Bunker buster, Smoke.
Firing bombardments		80"/200cm	-	5	2+	Smoke bombardment.

INFANTRY TEAMS

Team	Range	ROF	Anti-tank	Firepower	Notes
Pistol team	4"/10cm	1	1	6	
Rifle team	16"/40cm	1	2	6	
Rifle/MG team	16"/40cm	2	2	6	
SMG team	4"/10cm	3	1	6	Full ROF when moving.
Light Mortar team	16"/40cm	1	1	4+	Can fire over friendly teams.
Panzerschreck team	8"/20cm	2	11	5+	Tank Assault 5.
Flame-thrower team	4"/10cm	2	-	6	Flame-thrower.
Staff team	16"/40cm	1	2	6	Moves as a Heavy Gun team.

Additional Training and Equipment

Panzerfaust	4"/10cm	1	12	5+	Tank Assault 6, Cannot shoot in the Shooting Step if moved in the Movement Step.

Close-defence and Pioneer teams are rated as Tank Assault 4.

TRANSPORT TEAMS

Vehicle	Mobility	Armour Front	Side	Top	Equipment and Notes
Trucks					
3-ton truck	Wheeled	-	-	-	
Horse-drawn wagon	Wagon	-	-	-	
Tractors					
Stalinets tractor	Slow Tank	-	-	-	
Horse-drawn limber	Wagon	-	-	-	
Recovery Vehicles					
Famo	Half-tracked	-	-	-	Recovery vehicle.
ISU Recovery	Slow Tank	9	7	2	Recovery vehicle.

FORTIFICATIONS

Weapon	Mobility	Range	ROF	Anti-tank	Firepower	Notes
40mm Bofors AA Nest	Immobile	24"/60cm	4	6	4+	Anti-aircraft.
HMG Pillbox	Immobile	24"/60cm	6	2	6	ROF 3 when Pinned Down.
HMG Nest	Immobile	24"/60cm	6	2	6	ROF 3 when Pinned Down.

AIRCRAFT

Aircraft	Weapon	To Hit	Anti-tank	Firepower	Notes
Ju 87D Stuka	Bombs	4+	5	1+	
Ju 88	Bombs	4+	5	1+	

FRENCH

TASK FORCE A FFI FREE FRENCH

Fifteen Divisions

To reflect the variable quality of the French partisans, FFI companies are rated as Irregular. After deployment, but before the first turn, roll a die for each company and consult the Fifteen Divisions table to determine their Motivation and Skill characteristics.

Roll	Irregular
1	Confident Conscript
2	Confident Conscript
3	Fearless Conscript
4	Reluctant Trained
5	Reluctant Trained
6	Confident Trained

Centaine

Whenever a rule in the Flames Of War rulebook talks about a platoon, read that as an FFI Company.

Operation Jedburgh

An FFI Company with a Jedburgh Carbine team is not considered an Allied platoon.

Life in the Maquis

FFI Company teams are always Gone to Ground, unless they move At the Double, carry Passengers that are not Recce teams, shoot, or assault.

GOUM

Assault Troops

Goum Platoons may re-roll failed Motivation Tests to Rally from being Pinned Down.

Cautious Movement

Goum teams are always Gone to Ground, unless they move At the Double, carry Passengers that are not Recce teams, shoot, or assault.

Colonial Troops

Goum Platoons hit in assault combat on a roll of 3+.

No Obstacle Too Tough

Goum Infantry and Man-packed Gun teams are Mountaineers (see page 61 of the rulebook).

WEB INTELLIGENCE BRIEFINGS

FRENCH IN ITALY

Assault Troops

French Platoons may re-roll failed Motivation Tests to Rally from being Pinned Down.

It is a Matter of Honour

When Tirailleurs Platoons, Tirailleurs Ammunition & Pioneer Platoons, Tirailleurs Intelligence & Recon Platoons, and Engineer Combat Platoons test their Motivation to Counterattack in assaults, you may re-roll the die and apply the re-rolled result to all the above platoons that failed on the original roll.

Mountain Infantry

French Infantry and Man-packed Gun teams are Mountaineers (see page 61 of the rulebook).

Central Fire Control

French Company Command teams cannot act as spotting teams for artillery bombardments.

When a French Artillery battery that has a Staff team fires a bombardment, you may chose to group other artillery batteries with Staff teams into the same bombardment before rolling to Range In as a Central Fire Control Bombardment.

Use the Mixed Bombardments rule on page 131 of the rulebook.

FRENCH ARSENAL

TANK TEAMS

Name *Weapon*	Mobility *Range*	Front *ROF*	Armour Side *Anti-tank*	Top *Firepower*	Equipment and Notes
TANKS					
M5A1 Stuart	Light Tank	4	2	1	Co-ax MG, Hull MG, AA MG.
M6 37mm gun	*24"/60cm*	*2*	*7*	*4+*	*Stabiliser.*
M4 or M4A1 Sherman	Standard Tank	6	4	1	Co-ax MG, Hull MG, .50 cal AA MG.
M3 75mm gun	*32"/80cm*	*2*	*10*	*3+*	*Smoke, Stabiliser.*
SUPPORT WEAPONS					
M8 Scott HMC	Light Tank	3	2	0	.50 cal AA MG.
M1A1 75mm howitzer	*16"/40cm*	*2*	*6*	*3+*	*Smoke.*
Firing bombardments	*64"/160cm*	*-*	*3*	*6*	
TANK DESTROYERS					
M10 3in GMC	Standard Tank	4	2	0	.50 cal AA MG.
M7 3in gun	*32"/80cm*	*2*	*12*	*3+*	*Slow traverse.*
RECONNAISSANCE					
M3A1 armored car	Jeep	1	0	0	.50 cal AA MG, two AA MG.
Recon Jeep	Jeep	-	-	-	AA MG.
Mortar Jeep	Jeep	-	-	-	
M2 60mm Mortar	*24"/60cm*	*2*	*1*	*3+*	*Hull mounted, Portee, Can fire over friendly troops.*
Firing Bombardments	*32"/80cm*	*-*	*1*	*6*	
VEHICLE MACHINE-GUNS					
Vehicle MG	*16"/40cm*	*3*	*2*	*6*	*ROF 1 if other weapons fire.*
.50 cal Vehicle MG	*16"/40cm*	*3*	*4*	*5+*	*ROF 1 if other weapons fire.*

GUN TEAMS

Weapon	Mobility	Range	ROF	Anti-tank	Firepower	Notes
M2 .50 cal MG	Man-packed	16"/40cm	3	4	5+	
M1919 LMG team	Man-packed	16"/40cm	5	2	6	ROF 2 when pinned down or moving.
M1917 HMG team	Man-packed	24"/60cm	6	2	6	ROF 3 when pinned down or moving.
Captured HMG team	Man-packed	24"/60cm	6	2	6	ROF 3 when pinned down or moving.
M2 60mm mortar	Man-packed	24"/60cm	2	1	3+	Minimum range 8"/20cm.
Firing bombardments		32"/80cm	-	1	6	
M1 81mm mortar	Man-packed	24"/60cm	2	2	3+	Minimum range 8"/20cm, Smoke.
Firing bombardments		40"/100cm	-	2	6	Smoke bombardment.
Captured 81mm mortar	Man-packed	24"/60cm	2	2	3+	Minimum range 8"/20cm, Smoke.
Firing bombardments		40"/100cm	-	2	6	Smoke bombardment.
M1 Bofors gun	Immobile	24"/60cm	4	6	4+	Anti-aircraft, Turntable.
M3 37mm gun	Light	24"/60cm	3	7	4+	Gun shield.
Captured Anti-tank gun	Light	24"/60cm	3	8	4+	Gun shield.
M1 57mm gun	Medium	24"/60cm	3	10	4+	Gun shield, No HE.
M1A1 75mm pack howitzer	Light	16"/40cm	2	6	3+	Smoke.
Firing bombardments		56"/140cm	-	3	6+	Smoke bombardment.
M2A1 105mm howitzer	Immobile	24"/60cm	1	9	2+	Breakthrough gun, Gun shield, Smoke.
Firing bombardments		72"/180cm	-	4	4+	Smoke bombardment.
M1 155mm howitzer	Immobile	24"/60cm	1	10	1+	Bunker buster, Gun shield, Smoke.
Firing bombardments		88"/220cm	-	5	2+	Smoke bombardment.

INFANTRY TEAMS

Team	Range	ROF	Anti-tank	Firepower	Notes
Carbine team	8"/20cm	1	1	6	
Rifle team	16"/40cm	1	2	6	
MG team	16"/40cm	3	2	6	ROF 2 when pinned down.
SMG team	4"/10cm	3	1	6	Full ROF when moving.
PIAT or Bazooka team	8"/20cm	1	10	5+	Tank assault 4.
Staff team	16"/40cm	1	2	6	Moves as a Heavy Gun team.

ADDITIONAL TRAINING AND EQUIPMENT

Pioneer teams are rated as Tank Assault 3.

TRANSPORT TEAMS

Vehicle	Mobility	Armour Front	Armour Side	Armour Top	Equipment and Notes
TRUCKS					
Jeep	Jeep	-	-	-	Optional Passenger-fired AA MG or .50 cal AA MG.
Dodge ¾-ton, ½-ton or GMC 2½-ton truck	Wheeled	-	-	-	
ARMOURED PERSONNEL CARRIERS					
M2 or M3 half-track	Half-tracked	1	0	0	Optional Passenger-fired AA MG or .50 cal AA MG.

AIRCRAFT

Aircraft	Weapon	To Hit	Anti-tank	Firepower	Notes
P-47 Thunderbolt	MG	2+	6	5+	
	Bombs	4+	5	1+	
P-40 Warhawk	MG	2+	6	5+	
	Bombs	4+	5	2+	

HUNGARIAN

Huszár

Hungarian platoons that are Huszár Platoons are so indicated under those platoons.

Any Huszár Platoon with a Command team may attempt a Huszár move at the start of the Shooting Step instead of shooting. If a platoon attempts to make a Huszár move, it may not shoot even if it fails to make a Huszár move.

Roll a Skill test for each platoon:

- If the test is successful, the platoon may move another 4"/10cm,
- Otherwise the platoon cannot move this step.

All normal rules apply for this movement. Platoons cannot make Huszár moves if they are Pinned Down or have moved At the Double. Bogged Down or Bailed vehicles cannot make Huszár moves.

Preparing for the Coming Storm

Hungarian platoons may re-roll failed Skill Tests to Dig In.

Schürzen

When a tank that is protected by Schürzen is hit by a weapon with a Firepower of 5+ or 6 on the Side armour by shooting and fails its Armour Save, roll a special 4+ Schürzen save:

- If the save is successful the Schürzen protects the tank from the side shot.
- If the save is not successful the shot penetrated the side armour as normal.

ÖNKÉNTES PUSKÁS SZÁZAD

Local Knowledge

A Önkéntes Puskás Század may deploy a Hungarian platoon entirely made up of Infantry or Gun teams, or a mix of both, in Immediate Ambush (see page 267 of the rulebook) in addition to any Ambush deployment that may be allowed by the Mission played. The Immediate Ambushing platoon must be taken from the platoons that are to be deployed on table during deployment.

WEB INTELLIGENCE BRIEFINGS

HUSZÁRS

Horthy's Men

A Hungarian 1st Cavalry Division Company may re-roll failed Company Morale Checks.

HUNGARIAN INFANTRY

Guarding The Frontier

A Határvadász Század Always Defends in any Mission.

FÖHADNAGY ERVIN TARCZAY

Főhadnagy Ervin Tarczay is a Warrior Tank team and is rated Fearless Veteran.

Tarczay may join any Harckocsizó Század and replaces the Company Command Turan II or Panzer IV H tank for +40 points, or Tiger I E or Panther A tank for +75 points.

AROUND THE FLANKS

Tarczay and any platoon he has joined may choose to make a Stormtroopers move (see page 241 of the rulebook) instead of a normal Huszár move .

This means you can shoot in the Shooting Step as normal and choose to make a Stormtroopers move instead of assaulting in the Assault Step.

AS STRAIGHT AS AN ARROW

Tarczay re-rolls misses with his tank's main gun.

DESTROY THOSE ANTI-TANK GUNS!

Tarczay may re-roll any failed saves when hit by a Gun team in Defensive Fire during an Assault.

HUNGARIAN ARSENAL

TANK TEAMS

Name *Weapon*	Mobility *Range*	Armour Front *ROF*	Armour Side *Anti-tank*	Armour Top *Firepower*	Equipment and Notes
LIGHT TANKS					
Toldi II *20mm 36M anti-tank rifle*	Standard Tank *16"/40cm*	2 *3*	1 *5*	1 *5+*	Co-ax MG.
Toldi IIa *40mm 41M gun*	Standard Tank *24"/60cm*	2 *3*	1 *7*	1 *4+*	Co-ax MG.
TANKS					
Turán I *40mm 41M gun*	Standard Tank *24"/60cm*	5 *3*	3 *7*	1 *4+*	Co-ax MG, Hull MG, Protected ammo.
Turán II *75mm 41M gun*	Standard Tank *24"/60cm*	5 *2*	3 *9*	1 *3+*	Co-ax MG, Hull MG, Protected ammo.
Panzer IV H *7.5cm KwK40 gun*	Standard Tank *32"/80cm*	6 *2*	3 *11*	1 *3+*	Co-ax MG, Hull MG, Protected ammo, Schürzen.
Panther A *7.5cm KwK42 gun*	Standard Tank *32"/80cm*	10 *2*	5 *14*	1 *3+*	Co-ax MG, Hull MG, Wide tracks.
HEAVY TANKS					
Tiger I E *8.8cm KwK36 gun*	Slow Tank *40"/100cm*	9 *2*	8 *13*	2 *3+*	Co-ax MG, Hull MG, Protected ammo, Wide tracks. *Slow traverse.*
ASSAULT GUNS					
Zrínyi II *105mm 43M howitzer*	Standard Tank *24"/60cm*	7 *2*	2 *10*	1 *2+*	AA MG, Protected ammo. *Hull mounted, Breakthrough gun.*
StuG G *7.5cm StuK40 gun*	Standard Tank *32"/80cm*	7 *2*	3 *11*	1 *3+*	Hull MG, Protected ammo, Schürzen. *Hull mounted.*
Hetzer *7.5cm PaK39 gun*	Standard Tank *32"/80cm*	7 *2*	2 *11*	1 *3+*	Hull MG, Overloaded. *Hull mounted.*

Reconnaissance

Csaba 39M	Wheeled	1	0	0	Co-ax MG.
20mm 36M anti-tank rifle	*16"/40cm*	*3*	*5*	*5+*	
Fiat-Ansaldo 35M	Half-tracked	1	0	1	Twin MG, Unreliable.
Motorcycle Rifle/MG team	Jeep	-	-	-	Motorcycle reconnaissance, Dismount as Rifle/MG team.
Rifle/MG	*16"/40cm*	*2*	*2*	*6*	*Hull mounted, Vehicle MG.*
Motorcycle Panzerfaust SMG team	Jeep	-	-	-	Motorcycle reconnaissance, Dismount as Panzerfaust SMG team.
When firing as SMG	*4"/10cm*	*3*	*1*	*6*	*Hull-mounted, Vehicle MG.*
When firing as Panzerfaust	*4"/10cm*	*1*	*12*	*5+*	*Awkward layout.*

Self-propelled Anti-aircraft

Nimrod	Standard Tank	2	1	0	
40mm 36M Bofors gun	*24"/60cm*	*4*	*7*	*4+*	*Anti-aircraft.*

Vehicle Machine-guns

Vehicle MG	*16"/40cm*	*3*	*2*	*6*	*ROF 1 if other weapons fire.*
Vehicle Twin MG	*16"/40cm*	*4*	*2*	*6*	*ROF 2 if other weapons fire.*

GUN TEAMS

Weapon	Mobility	Range	ROF	Anti-tank	Firepower	Notes
7/31M HMG	Man-packed	24"/60cm	6	2	6	ROF 3 when pinned down or moving
20mm 36M anti-tank rifle	Man-packed	16"/40cm	3	5	5+	
81mm 36/39M mortar	Man-packed	24"/60cm	2	2	3+	Smoke, Minimum range 8"/20cm.
Firing bombardments		40"/100cm	-	2	6	Smoke bombardment.
120mm 43M mortar	Light	56"/140cm	-	3	3+	
37mm 36M gun	Light	24"/60cm	3	6	4+	Gun shield.
Firing Stielgranate		8"/20cm	1	12	5+	
40mm 40M gun	Light	24"/60cm	3	7	4+	Gun shield.
Firing Stielgranate		8"/20cm	1	12	5+	
75mm 40M gun (7.5cm PaK40)	Medium	32"/80cm	2	12	3+	Gun shield.
40mm 36M Bofors gun	Immobile	24"/60cm	4	7	4+	Anti-aircraft, Turntable.
80mm 29/38M Bofors gun	Immobile	32"/80cm	2	12	3+	Heavy anti-aircraft, Turntable.
8.8cm FlaK37 gun	Immobile	40"/100cm	2	13	3+	Gun shield, Heavy anti-aircraft, Turntable.
75mm 15M howitzer	Medium	16"/40cm	2	5	3+	Gun shield.
Firing bombardments		64"/160cm	-	3	6	
100mm 14M howitzer (100/17)	Immobile	24"/60cm	1	9	2+	Breakthrough gun, Gun shield.
Firing bombardments		72"/180cm	-	4	4+	
105mm 37M howitzer (leFH18)	Immobile	24"/60cm	1	10	2+	Breakthrough gun, Gun shield, Smoke.
Firing bombardments		72"/180cm	-	4	4+	Smoke bombardment.
149mm 14/31M howitzer	Immobile	16"/40cm	1	8	1+	Bunker buster, Gun shield.
Firing bombardments		72"/180cm	-	5	2+	
150mm 43M rocket launcher	Light	64"/160cm	-	3	4+	Rocket launcher, Smoke bombardment.

FORTIFICATIONS

Weapon	Mobility	Range	ROF	Anti-tank	Firepower	
40mm Bofors AA Nest	Immobile	24"/60cm	4	7	4+	Anti-aircraft.
HMG Pillbox	Immobile	24"/60cm	6	2	6	ROF 3 when Pinned Down.
HMG Nest	Immobile	24"/60cm	6	2	6	ROF 3 when Pinned Down.

INFANTRY TEAMS

Team	Range	ROF	Anti-tank	Firepower	Notes
Rifle team	16"/40cm	1	2	6	
Rifle/MG team	16"/40cm	2	2	6	
SMG team	4"/10cm	3	1	6	Full ROF when moving.
Light Mortar team	16"/40cm	1	1	4+	Can fire over friendly teams.
Panzerschreck team	8"/20cm	2	11	5+	Tank Assault 5.
Flame-thrower team	4"/10cm	2	-	6	Flame-thrower.
Staff team	16"/40cm	1	2	6	Moves as a Heavy Gun team.
Additional Training and Equipment					
Panzerfaust	4"/10cm	1	12	5+	Tank Assault 6, Cannot shoot in the Shooting Step if moved in the Movement Step

Páncélvadész and Pioneer teams are rated as Tank Assault 4.

TRANSPORT TEAMS

Vehicle	Mobility	Armour Front	Armour Side	Armour Top	Equipment and Notes
Trucks					
Motorcycle team or Field Car	Jeep	-	-	-	
Botond, 3-ton, Kfz 68 radio, or Kfz 70 truck	Wheeled	-	-	-	
Horse-drawn wagon	Wagon	-	-	-	
Tractors					
Horse-drawn limber	Wagon	-	-	-	
Sd Kfz 11 tractor	Half-tracked	-	-	-	
Recovery Vehicles					
Famo half-track	Half-tracked	-	-	-	Recovery vehicle.

AIRCRAFT

Aircraft	Weapon	To Hit	Anti-tank	Firepower	Notes
Ju 87D Stuka	Bombs	4+	5	1+	
Ju 87G Stuka	Cannon	3+	11	4+	

POLISH

Fate of the Nation
A Polish Platoon joined by a 2iC team may re-roll failed Morale Checks.

ARMII KRAJOWEJ

Home-built Weapons
Weapons that are listed in the Arsenal as Home-built are liable to explode with deadly result.

After a platoon with Home-built weapons has resolved all of its shooting in the step, roll a die for each Home-built weapon that shot:

- On a roll of 1 the team shooting the Home-built weapon is Destroyed.
- Otherwise, the weapon holds up and the team is not Destroyed.

For example, when a Home-built Flame-thrower rolls a 1 the Flame-thrower team is Destroyed rather than removed from play after it is done shooting.

Battalions of the Uprising
A Batalionowy Armia Krajowa uses the Soviet Centralised Control special rule found on page 250 of the rulebook.

Tunnels and Paths
Infantry and Man-packed Gun teams from Combat and Weapons Companies and the Kedyw Assault Platoon treat rubble and woods as Cross-country Terrain, allowing them to move At The Double.

1ST POLISH INDEPENDENT PARACHUTE BRIGADE

Polish Parachute Companies use all of the normal British special rules found on pages 246 to 248 of the rulebook, except British bulldog.

For Your Freedom and Ours
Polish Platoons fight to the last and may re-roll failed Platoon Morale Checks. If you are fielding a Polish Company, your Company Command team may re-roll Company Morale Checks.

POLISH 1ST ARMOURED DIVISION & POLES IN ITALY

Polish forces from these PDFs use all of the normal British special rules found on pages 246 to 248 of the rulebook, except British bulldog. In additons they use the Fate of the Nation special rule above.

MAJOR GENERAL STANISLAW SOSABOWSKI

Major General Stanislaw Sosabowski is a Warrior Higher Command Rifle team rated Fearless Veteran. Major General Stanislaw Sosabowski may join a Polish Parachute Company (found in *A Bridge Too Far*) for +45 points.

What About the Germans?

At the start of the game after both players have deployed, but before any Reconnaissance Deployment moves, a force commanded by Sosabowski may redeploy a single platoon anywhere within its normal deployment area.

If the platoon was in Prepared Positions at the start of the game, they remain in Prepared Positions after this redeployment.

Timely Reinforcements

Once per turn, a force led by Sosabowski may re-roll one die rolled to receive Reserves.

WOJTEK THE SOLDIER BEAR

Wojtek the Soldier Bear is a Warrior Transport team rated Fearless Trained. Wojtek may join a Polish Kompania Piechorty for +70 points. Wojtek the Soldier Bear is not part of any platoon.

Supply Carrier

During your turn a Field Artillery Battery can use Wojtek the Soldier Bear to allow it to deliver a burst of fire in a Shooting Step. Wojtek must be In Command with the battery. Wojtek is removed from the table and placed in Reserve before any dice are rolled.

A battery delivering a burst of fire can either:

- Count as having twice as many weapons when firing an Artillery bombardment in this Shooting Step, or
- Roll one extra die to hit for each team that is In Command when shooting in this Step.

Ammuntion Bear-er

While Wojtek is off of the table roll a die at the beginning of your turn when you would normally roll for reserves.

- On a result of 5+, Wojtek arrives from Reserve.
- On any other result, Wojtek his team are still gathering more ammuntion and do not arrive this turn.

Lovable Mascot

While Wojtek is on the table he may attempt to inspire the men around him. Wojtek is removed from the table and placed in Reserve before any dice are rolled to re-roll one failed Platoon Morale Check of one Polish platoon that is within Command Distance of Wojtek.

POLISH ARSENAL

TANK TEAMS

Name *Weapon*	Mobility *Range*	Front *ROF*	Armour Side *Anti-tank*	Top *Firepower*	Equipment and Notes
TANKS					
Panzer IV	Standard Tank	6	3	1	Co-ax MG, Hull MG, Protected Ammo, Schürzen, Unreliable.
7.5cm KwK40 gun	*32"/80cm*	*1*	*11*	*3+*	
Panther A	Standard Tank	10	5	1	Co-ax MG, Hull MG, Wide Tracks, Unreliable.
7.5cm Kwk42 gun	*32"/80cm*	*1*	*14*	*3+*	
Tiger I E	Slow Tank	9	8	2	Co-ax MG, Hull MG, Protected Ammo, Wide tracks, Unreliable.
8.8cm KwK36 gun	*40"/100cm*	*1*	*13*	*3+*	*Slow Traverse.*
Hetzer	Standard Tank	7	2	1	Hull MG, Overloaded, Unreliable.
7.5cm PaK39 gun	*32"/80cm*	*1*	*11*	*3+*	*Hull mounted.*
ARMOURED CARS					
Sd Kfz 221 (MG)	Wheeled	1	0	0	AA MG, Unreliable.
VEHICLE MACHINE-GUNS					
Vehicle MG	*16"/40cm*	*3*	*2*	*6*	*ROF 1 if other weapons fire.*

GUN TEAMS

Weapon	Mobility	Range	ROF	Anti-tank	Firepower	Notes
PTRD anti-tank rifle	Man-packed	16"/40cm	2	5	5+	Tank Assault 3.
MG34, MG42 or Maksim HMG	Man-packed	24"/60cm	6	2	6	ROF 3 when pinned down or moving
Home-built 81mm mortar	Man-packed	24"/60cm	2	2	3+	Home-built, Minimum range 8"/20cm.
Firing bombardments		32"/80cm	-	2	6	
8cm GW34 mortar	Man-packed	24"/60cm	2	2	3+	Smoke, Minimum range 8"/20cm.
Firing bombardments		40"/100cm	-	2	6	Smoke bombardment.
82-BM-41 mortar	Man-packed	24"/60cm	2	2	3+	Minimum range 8"/20cm.
Firing bombardments		40"/100cm	-	2	6	
3.7cm PaK36 gun	Light	24"/60cm	3	6	4+	Gun shield.
Firing Stielgranate		8"/20cm	1	12	5+	
45mm obr 1942 gun	Light	24"/60cm	3	8	4+	Gun shield.
5cm PaK38 gun	Medium	24"/60cm	3	9	4+	Gun shield.
7.5cm PaK40 gun	Medium	32"/80cm	2	12	3+	Gun shield.
10.5cm leFH18 howitzer	Immobile	24"/60cm	1	10	2+	Breakthrough gun, Gun shield, Smoke.
Firing bombardments		72"/180cm	-	4	4+	Smoke bombardment.

INFANTRY TEAMS

Team	Range	ROF	Anti-tank	Firepower	Notes
Rifle team	16"/40cm	1	2	6	
Rifle/MG team	16"/40cm	2	2	6	
SMG team	4"/10cm	3	1	6	Full ROF when moving.
Bazooka or PIAT team	8"/20cm	1	10	5+	Tank assault 4.
Flame-thrower team	4"/10cm	4	-	6	Flame-thrower. Home-built.
Staff team	16"/40cm	1	2	6	Moves as a Heavy Gun team.

Additional Training and Equipment

Team	Range	ROF	Anti-tank	Firepower	Notes
Panzerfaust	4"/10cm	1	12	5	Tank Assault 6, Cannot shoot in the Shooting Step if moved in the Movement Step.

Pioneer teams are rated as Tank Assault 4. Teams with Gammon Bombs are rated as Tank Assault 3.

TRANSPORTS TEAMS

Name *Weapon*	Mobility *Range*	Armour Front *ROF*	Side *Anti-tank*	Top *Firepower*	Equipment and Notes
Armoured Personnel Carriers					
Sd Kfz 251/1 half-track	Half-tracked	1	0	0	Hull MG, Passenger-fired AAMG, Unreliable.
Kubus	Wheeled	1	0	0	Co-ax MG, Overloaded, Unreliable.
Home-built Flame-thrower	*4"/10cm*	*3*	*-*	*6*	*Flame-thrower. Home-built.*

ROMANIAN

Peasant Army

After deployment, but before the first turn, roll a die for each Romanian platoon, Independent team, and Warrior team. Consult the Peasant Army table to determine their Skill and Motivation ratings. Platoons held in Reserves do not roll until they arrive.

Peasant Army Table

Roll	Regular	Elite
1	Reluctant Trained	Reluctant Veteran
2	Reluctant Veteran	Reluctant Veteran
3	Confident Trained	Confident Trained
4	Confident Trained	Confident Trained
5	Confident Trained	Confident Veteran
6	Confident Veteran	Confident Veteran

French Doctrine

Romanian Infantry and Gun teams attempting to Dig In succeed on a roll of 3+.

When conducting Defensive Fire, Infantry and Gun teams that are Dug In or in Entrenchments can shoot over any Man-packed or Light Gun teams that are also Dug In or in Entrenchments.

Central Fire Control

Romanian Company Command teams cannot act as spotting teams for Artillery Bombardments.

When a Romanian artillery battery that has a Staff team fires an Artillery Bombardment, any other artillery batteries with Staff teams may join it before rolling to Range In and fire as a single combined Artillery Battery.

Hated Enemy

When Romanian platoons test their Motivation to Counterattack in assaults, you may re-roll the die and apply the re-rolled result to all Romanian platoons that failed on the original roll.

Soviet Battalion Komissar

A Soviet Battalion Komissar team in a Romanian Battalion HQ is a Warrior team.

If a Soviet Battalion Komissar team joins a Romanian company or platoon in the Shooting or Assault Step before the company or platoon takes any hits from shooting, they must attempt to sustain the Romanians' morale.

If the company or platoon takes five of more hits from shooting, roll a die after all shooting at the platoon is completed.

On a roll of 2+ you must immediately Destroy any other Infantry or Gun team from the company or platoon within Command Distance of the Komissar team.

Otherwise on roll of 1, the Romanians rebel and shoot the Komissar team, which is Destroyed.

If the Komissar team survived the attempt to sustain the Romanians' morale, the company or platoon, if it started the step with at least fifteen Infantry or Gun teams (including the Komissar team and any attached teams) In Command, now must take at least ten hits in a single Shooting Step before they become Pinned Down.

A single hit from an Artillery bombardment still pins them, however.

A Soviet Battalion Komissar may attach to Allied Soviet Companies and uses the normal Battalion Komissar rules when they do so, see page 249 of the rulebook.

One-man Turret

Any tank with a One-man Turret that moves adds +1 to the score required to hit when shooting with its turret main gun. Machine-guns and Flame-throwers do not suffer this penalty.

In addition, a tank with a One-man Turret may not fire both the turret main gun and the turret machine-gun in the same turn.

BATALION PUSCASI

Stalin's Cannon Fodder

A Soviet allied Batalion Puscasi Always Attacks against any other Infantry Company in a mission with the Prepared Positions special rule (see page 264 of the rulebook).

Eastern Legions

While playing a Batalion Puscasi, whenever the rules talk about a Platoon, read that as a Puscasi Company. Where the rules talk about a Company, read that as a Batalion Puscasi.

SERGENT AUREL SANDU

Sergent Aurel Sandu is a Warrior Pioneer Rifle/MG team. He replaces a Pioneer Rifle/MG team in a Puscasi Pioneer Platoon or a Pioneer Platoon for +45 points.

He served through 1944 and 1945 and can be taken with either a German allied or Soviet allied Romanian force.

Lucky Aurel

Once per game Sandu may re-roll a failed Infantry Save.

Old Hand

Sandu passes all Skill tests on a 3+.

Lending a Hand

One Gun team within 4"/10cm of Sandu may re-roll failed To Hit rolls if Sandu does not shoot in the Shooting step.

Unstable Mines

If Sandu's platoon's Pioneer Supply vehicle is used to place a Minefield in a Mission with the Prepared Positions special rule, you may choose to place two Unstable Minefields instead.

When you place the Pioneer Supply vehicle, instead of placing one Minefield, place two Unstable Minefields. Roll a dice for each Unstable Minefield you placed. On a 2+ they are placed as normal.

On a roll of 1 the Unstable Minefield explodes while it is being placed. The Unstable Minefield is removed from the table and all the teams in Sandu's Pioneer Platoon must make an immediate Infantry save. Teams that fail the Infantry save are removed from the table, but do not count as casualties for the purposes of Platoon Morale

Unstable Minefields count as Minefields for all other purposes.

ROMANIAN ARSENAL

TANK TEAMS

Name *Weapon*	Mobility *Range*	Front *ROF*	Armour Side *Anti-tank*	Top *Firepower*	Equipment and Notes
LIGHT TANKS					
R-2 (Panzer 35(t))	Standard Tank	2	1	1	Co-ax MG, Hull MG, Protected ammo, Unreliable.
3.7cm KwK35(t) gun	*24"/60cm*	*2*	*6*	*4+*	
R-35	Slow Tank	3	3	1	Co-ax MG.
37mm SA18 gun	*16"/40cm*	*2*	*4*	*4+*	*One-man turret.*
R-35 Vanatorul de Care	Slow Tank	3	3	1	
45mm obr 1938 gun	*24"/60cm*	*2*	*7*	*4+*	*One-man turret.*
T-38 (Panzer 38(t))	Standard Tank	2	1	1	Co-ax MG, Hull MG, Protected ammo, Unreliable.
3.7cm KwK35(t) gun	*24"/60cm*	*2*	*6*	*4+*	
MEDIUM TANKS					
T-3 (Panzer III N)	Standard Tank	6	3	1	Co-ax MG, Hull MG, Protected ammo.
7.5cm KwK36 gun	*24"/60cm*	*2*	*9*	*3+*	
T-4 (Panzer IV H)	Standard Tank	6	3	1	Co-ax MG, Hull MG, Protected ammo, Schürzen.
7.5cm KwK40 gun	*32"/80cm*	*2*	*11*	*3+*	
ASSAULT GUNS					
TA (StuG G)	Standard Tank	7	3	1	Hull MG, Protected ammo, Schürzen.
7.5cm StuK40 gun	*32"/80cm*	*2*	*11*	*3+*	*Hull mounted.*
SELF PROPELLED ANTI TANK					
TACAM T-60	Standard Tank	1	0	0	AA MG, Overloaded.
76mm F-22 gun	*32"/80cm*	*2*	*10*	*3+*	*Hull mounted.*
TACAM R-2	Standard Tank	1	0	0	Hull MG, Unreliable.
76.2mm ZIS-3 gun	*32"/80cm*	*2*	*9*	*3+*	*Hull mounted.*
ARMOURED CARS					
AB (Sd Kfz 222)	Wheeled	1	0	0	Co-ax MG.
2cm KwK38 gun	*16"/40cm*	*3*	*5*	*5+*	*Self-defence anti-aircraft.*
VEHICLE MACHINE-GUNS					
Vehicle MG	*16"/40cm*	*3*	*2*	*6*	*ROF 1 if other weapons fire.*

GUN TEAMS

Weapon	Mobility	Range	ROF	Anti-tank	Firepower	Notes
ZB 53 HMG	Man-packed	24"/60cm	6	2	6	ROF 3 when pinned down or moving
60 mm Brandt mortar	Man-packed	24"/60cm	2	1	3+	Can fire over friendly troops.
Firing bombardments		32"/80cm	-	1	6	
81.4mm Brandt mortar	Man-packed	24"/60cm	2	2	3+	Smoke, Minimum range 8"/20cm.
Firing bombardments		40"/100cm	-	2	6	Smoke bombardment.
120mm Reşiţa mortar	Light	56"/140cm	-	3	3+	
Mitral 20mm gun (FlaK38)	Light	16"/40cm	4	5	5+	Anti-aircraft, Gun shield, Turntable.
TAC 37mm gun (3.7cm PaK36)	Light	24"/60cm	3	6	4+	Gun shield.
TAC 37 45mm gun (obr 1937)	Light	24"/60cm	3	7	4+	Gun shield.
TAC 42 45mm gun (obr 1942)	Light	24"/60cm	3	8	4+	Gun shield.
TAC 47mm Bohler gun (47/32)	Man-packed	24"/60cm	3	7	4+	
TAC 47mm Schneider gun	Light	24"/60cm	3	8	4+	Gun shield.
TAC 38 50mm gun (PaK38)	Medium	24"/60cm	3	9	4+	Gun shield.
TAC 938 75mm gun (PaK38/97)	Medium	24"/60cm	2	10	3+	Gun shield.
TAC 40 75mm gun (PaK40)	Medium	32"/80cm	2	12	3+	Gun shield.
TAC 43 Reşiţa 75mm gun	Heavy	32"/80cm	2	13	3+	Gun shield.
KF 75mm gun (75/27)	Heavy	24"/60cm	2	8	3+	Gun shield, Smoke.
Firing bombardments		64"/160cm	-	3	6	Smoke bombardment.
RF 75mm gun (76K/02)	Heavy	24"/60cm	2	8	3+	Gun shield, Smoke.
Firing bombardments		64"/160cm	-	3	6	Smoke bombardment.
Skoda 100mm M14 howitzer (100/17)	Immobile	24"/60cm	1	9	2+	Breakthrough gun, Gun shield.
Firing bombardments		72"/180cm	-	4	4+	
Skoda 100mm M30 howitzer (leFH30)	Immobile	24"/60cm	1	9	2+	Breakthrough gun, Gun shield.
Firing bombardments		88"/220cm	-	4	4+	
Schneider 105mm M36 gun	Immobile	32"/80cm	1	12	2+	Breakthrough gun, Gun shield.
Firing bombardments		88"/220cm	-	4	4+	
Skoda 150mm M34 howitzer	Immobile	32"/80cm	1	13	1+	Bunker buster, Gun shield.
Firing bombardments		88"/220cm	-	5	2+	
15cm sFH18 howitzer	Immobile	24"/60cm	1	13	1+	Bunker buster, Smoke.
Firing bombardments		80"/200cm	-	5	2+	Smoke bombardment.

INFANTRY TEAMS

Team	Range	ROF	Anti-tank	Firepower	Notes
Rifle team	16"/40cm	1	2	6	
Rifle/MG team	16"/40cm	2	2	6	
MG team	16"/40cm	3	2	6	ROF 2 when pinned down
SMG team	4"/10cm	3	1	6	Full ROF when moving.
Panzerschreck team	8"/20cm	2	11	5+	Tank Assault 5.
Flame-thrower team	4"/10cm	2	-	6	Flame-thrower.
Staff team	16"/40cm	1	2	6	Moves as a Heavy Gun team.

Additional Training and Equipment

Team	Range	ROF	Anti-tank	Firepower	Notes
Panzerfaust	4"/10cm	1	12	5+	Tank Assault 6, Cannot shoot in the Shooting Step if moved in the Movement Step.

Pioneer teams are rated as Tank Assault 4.

TRANSPORT TEAMS

Vehicle Weapon	Mobility Range	Armour Front ROF	Side Anti-tank	Top Firepower	Equipment and Notes
TRUCKS					
Kfz 15 field car or Kübelwagen jeep	Jeep	-	-	-	
1.5-ton, 3-ton, Kfz 70, or Kfz 68 Radio truck	Wheeled	-	-	-	
Horse-drawn wagon or Cavalry wagon	Wagon	-	-	-	
TRACTORS					
RSO tractor	Slow Tank	-	-	-	
Sd Kfz 7 and 11 half-tracks	Half-tracked	-	-	-	
Horse-drawn limber or Cavalry limber	Wagon	-	-	-	
ARMOURED PERSONNEL CARRIERS					
TB (Sd Kfz 251 D)	Half-tracked	1	0	0	Hull MG, Passenger-fired AA MG.

AIRCRAFT

Aircraft	Weapon	To Hit	Anti-tank	Firepower	Notes
Ju 87D Stuka	Bombs	4+	5	1+	
Ju 87G Stuka	Cannon	3+	11	4+	
Hs 129B	Cannon	2+	9	4+	Flying Tank.

CLARIFICATIONS

This section clarifies issues that have arisen in current compilations and intelligence briefings. These books may also have changes to their warriors, special rules, and arsenals in addition to the clarifications noted here. In particular, note that many guns between 100mm and 128mm have gained the Breakthrough Guns special rule, some unarmoured self-propelled guns have gained the Gun Shields special rule, and US armoured anti-aircraft half-tracks no longer have the Awkward Layout special rule, although their weapons are now rated as guns rather than machine-guns.

RED BEAR

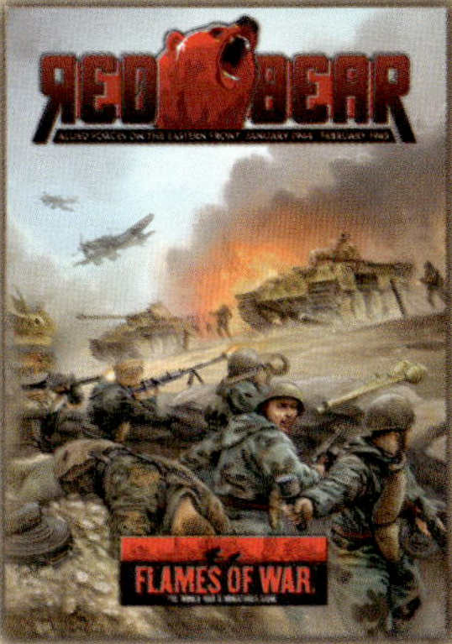

Peredovoye Otryad (page 76)

A Peredovoye Otryad should be a Mechanised Company, not a Tank Company.

Razvedki Anti-tank Company (page 81)

The 76mm ZIS-3anti-tank gun should be shown with four crew.

Inomarochnikiy Tankovy Company (page 126)

The points value for 6 M4 76mm (M4A2 Sherman) tanks should be 545 points.

Arsenal (pages 148 to 151)

The entry labelled Mark VIII (Valentine VIII) should be Mark III (Valentine VIII). The Tachanka has the characteristics shown on page 88. The Captured Sd Kfz 250 half-track has the characteristics shown on page 89. The 122mm A-19 obr 1931/37 gun should have a bombardment range of 96"/240cm. The Maksim HMG Nest and 45mm obr 1942 Nest have the characteristics shown on page 89.

AK Machine-gun Company (page 167)

The choices should be 6 MG42 or Ckm wz.30 and 3 MG42 or Ckm wz.30.

Arsenal (page 170)

The 10.5cm leFH18 howitzer should have a bombardment range of 72"/180cm.

Puscasi Anti-tank Company (page 194)

The TAC 37mm should read TAC 37 45mm.

Stalin's Europe, River of Heroes, Hammer and Sickle, and Stalin's Onslaught

Due to the major changes to Hen and Chicks in Version 3 these books are no longer valid for tournament use.

GREY WOLF

Fallschirmjäger Platoon (page 33)

The option should read: Replace Command Rifle/MG team with a Command Panzerknacker SMG team for +5 points or with a Command Panzerfaust SMG team for +10 points.

SS-Freiwilligen-Panzergrenadierkompanie (page 38), SS-Panzerkampfgruppe (page 108), SS-Panzergrenadierkampfgruppe (page 110)

The second sentence of the organisation box should start: "Your Company HQ, Combat, Weapons, and Support platoons must be…".

Sperrverband (page 95)

The rule box under the Sperrverband HQ is incorrect. The correct rule is shown on page 33.

Strongpoints (pages 158 to 159)

All Field Strongpoints, Street Strongpoints, and Gun Positions must be attached to a platoon.

Gepanzerte SS-Panzerpionier Platoon (page 177)

This platoon is missing the rule box saying: A Gepanzerte SS-Panzerpionier Platoon may use the Mounted Assault special rule.

Arsenal (pages 192 to 193)

The first entry labelled Panzer III OP should read Panzer II OP. The 10cm leFH14/19 (p) howitzers of the Heavy Artillery Car should have Smoke on their direct fire line. The armoured train's Panzer 38(t) tank should be a Panzer 38(t) S tank.

Pioneeri Platoon (page 251)

The second option should read: Replace all Pioneer Rifle teams in Pioneeri Squads with Pioneer SMG teams for +10 points per Pioneeri Squad.

Turning Tide

Carrier Platoon (page 48), Scout Platoon (page 104), Lorried Carrier Platoon (page 111)

Each Scout or Carrier Patrol operates as a separate platoon for all purposes.

Commando HQ (page 66)

PIAT Rifle teams in the Commando HQ should be PIAT teams.

Commando Company (page 66)

Each Commando Section operates as a separate platoon for all purposes. You may only replace rifle teams with flame-throwers in one section in one Commando Company in the entire Commando.

SAR Recce Platoon (page 98)

Ignore the rule box starting 'Recce Patrols operate…'.

Naval Gunfire Support (page 119)

The rule should read: If you have Naval Gunfire Support, your force will field an Observer Rifle team that can spot for an artillery battery of four Confident Trained Naval Guns with no Staff team. The guns are not deployed on the table, but have the range to hit any target on the table.

Arsenal (page 125)

The Churchill Crocodile should not have Semi-indirect Fire. A Churchill AVRE has a Hull MG. The Humber LRC III has Front Armour 0.

British National Rules (pages 128 to 129)

The British national rules in the rulebook replace those shown here.

Airborne Divisional Recon Platoon (page 147)

Recon and Support Sections operate as separate platoons for all purposes.

899th Tank Destroyer Platoon (page 149), Task Force Tank Destroyer Platoon (page 228), Tank Destroyer Platoon (page 231)

Replace the Command Carbine team and Jeep with a .50 cal Recon Jeep and each other Carbine team and M20 utility with an M20 scout car. The jeep and scout cars form the Security Section. The characteristics for these are on page 21.

Machine-gun Platoon (page 174)

A Machine-gun Platoon can Combat attach to Rifle Platoons (on page 219).

Cavalry Recon Platoon (page 178), Task Force Cavalry Recon Platoon (page 227)

Cavalry Recon Patrols operate as separate platoons for all purposes.

Ranger Company (page 188)

Each Ranger Platoon operates as a separate platoon for all purposes.

Armored Anti-tank Platoon (page 208)

The cost for 2 Gun Sections for the 2nd Armored Division should be 80 points.

Company HQ (page 211, 227)

The M8 armoured cars of the Company HQ are Recce teams.

Armored Recon Platoon (page 211)

Armored Recon Patrols operate as separate platoons for all purposes. An Armoured Recon Patrol is a Reconnaissance Platoon.

Towed Tank Destroyer Platoon (page 230)

The Command Carbine team is part of the Tank Destroyer Section. Replace the Jeep with .50 cal AA MG with a Command .50 cal Recon Jeep and each other Carbine team and Jeep with a Recon Jeep. These jeeps form the Security Section.

Corps Field Artillery Battery (page 232)

The first rule box should read: You may not field a Corps Field Artillery Battery (155mm) unless you are fielding a Field Artillery Battery or Corps Field Artillery Battery with at least as many M2A1 105mm howitzers.

Naval Gunfire Support (page 119)

The rule should read: If you have Naval Gunfire Support, your force will field an Observer Carbine team that can spot for an artillery battery of four Confident Trained Naval Guns with no Staff team. The guns are not deployed on the table, but have the range to hit any target on the table.

Arsenal (pages 237 to 239)

M2 37mm half-track has a Passenger-fired .50 cal AA MG and the 37mm is hull mounted.

US National Rules (pages 240 to 241)

The US national rules in the rulebook replace those shown here.

EARTH AND STEEL

Fortifications (pages 32 to 37)

The Fortifications rules in the rulebook replace those shown here.

Panzerpionierkompanie HQ (page 67)

The option should read: Replace Command SMG teams with Command Panzerfaust SMG teams for +10 points per team.

Panzerpionier Platoon (page 67)

The first option should read: Replace Command Pioneer MG team with Command Pioneer Panzerfaust SMG team for +10 points.

Luftwaffe Pionier Platoon (page 86)

The first option should read: Replace Command Pioneer Rifle team with Command Pioneer Panzerknacker SMG team for +5 points.

Schwere Panzer Pioneer Platoon (page 92), Panzer Lehr Pioneer Platoon (page 100), SS-Panzer Pioneer Platoon (153)

The option should read: Replace all Maultier half-tracks with Sd Kfz 251/7 (Pioneer) half-tracks for +10 points per half-track.

Panzergrenadier Lehr Anti-aircraft Platoon (page 106)

The platoon is equipped with armoured Sd Kfz 10/5 (2cm) half-tracks. These have Front, Side, and Top armour ratings of 0.

SS-Panzer Anti-aircraft Gun Platoon (page 152)

Any Sd Kfz 7/1 (Quad 2cm) half-tracks equipping the platoon is equipped with armoured half-tracks. These have Front, Side, and Top armour ratings of 0.

Arsenal (pages 206 to 209)

The range for the 2cm FlaK38 (V) on the Panzer IV chassis should be 16"/40cm, not 17"/40cm. The Sd Kfz 250 (Recce) and Sd Kfz 251/1 (Stuka) half-tracks should both have a Hull MG and an AA MG. The Jagdpanther should be called a Jagdpanther (early) to distinguish it from the later production model which wasn't Unreliable.

Special Rules (pages 210 to 213)

The national special rules in the rulebook replace those shown here. The Armoured Car Patrols rule becomes: Panzerspäh Patrols operate as separate platoons, each with their own command team.

HELL'S HIGHWAY

GUARDS RECCE PATROL (PAGE 18)

The option should read: Arm any or all Stuart tanks with AA MG for + 5 points per tank.

SQUADRON HQ (PAGE 31)

The second option should read: Arm any or all Staghound armoured cars with AA MG for + 5 points per tank.

HEAVY ARMOURED CAR PLATOON (PAGE 32)

The platoon choice should read 2 AEC II and 1 Daimler Dingo.

FIELD BATTERY, ROYAL ARTILLERY (PAGE 34). FIELD BATTERY (SP), ROYAL ARTILLERY (PAGE 35). MEDIUM BATTERY, ROYAL ARTILLERY (PAGE 36)

Each Gun Troop operates as a separate platoon for all purposes. The Command team and the Staff team from the HQ Troop must be attached to Gun troops at the start of the game (see Independent Command on page 248 of the rulebook).

AIR OBSERVATION POST (PAGES 39, 49)

The Air Observation Post rules on page 139 of the rulebook replace the Liaison Aircraft, Anti-aircraft Fire, and Air Observation Post rules. The US AOP does not use the Column Security special rule.

PARACHUTE RIFLE COMPANY (PAGE 42)

Your company may only have support options from either the 82nd Support Platoons or the 101st Support Platoons section, but not both.

PARACHUTE RIFLE PLATOON (PAGE 43)

The Mortar Squad is included in all platoons along with the HQ Section.

FALLSCHIRMJÄGERKOMPANIE (PAGE 64)

Your company may only have support options from either the 3. FJ Division Support Platoons or the 6. FJ Division Support Platoons section, but not both.

A BRIDGE TOO FAR

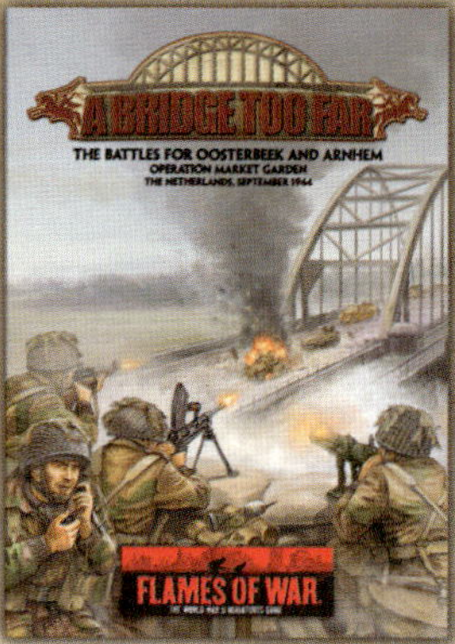

AIRLANDING LIGHT BATTERY (PAGE 27)

Each Gun Troop operates as a separate platoon for all purposes. The Command team and the Staff team from the HQ Troop must be attached to Gun troops at the start of the game (see Independent Command on page 248 of the rulebook).

REINFORCED TRUCK (PAGE 48)

The rule box should read: If a Reinforced Truck is hit, Infantry teams carried as passengers survive their Passenger Save on a roll of 3+.

GRENADIER COMPANY (PAGE 66)

The first option in both types of company should read: Replace any or all Command Rifle teams with Command Panzerknacker Rifle teams for +5 points per team.

ARSENAL (PAGE 76)

The Königstiger (Henschel) should not be Unreliable. The M5 37mm gun of the Humber IV is not Self-defence Anti-aircraft.